FM 21-50

FIELD MANUAL } HEADQUARTERS,
No. 21-50 } DEPARTMENT OF THE ARMY
WASHINGTON 25, D. C., *20 August 1957*

RANGER TRAINING

	Paragraph	Page
CHAPTER 1. INTRODUCTION		
Purpose and scope	1	3
Concepts of ranger training	2	3
2. RANGER TYPE OPERATIONS		
Section I. General		
Definition	3	5
Requisites for success	4	5
II. Aircraft support of ranger type operations		
Tactical use	5	5
Operations against infiltration and guerilla action	6	7
Night operations	7	7
Raid	8	8
Patrolling	9	9
Operations in special area	10	10
CHAPTER 3. AMBUSH AND ROADBLOCK TECHNIQUES		
Introduction	11	17
Conduct of an ambush	12	18
Establishing a roadblock	13	22
Defense against ambush	14	24
4. CLIFF ASSAULT TECHNIQUES		
Purpose and scope	15	30
Introduction	16	30
Special equipment	17	31
Initial landing	18	31
Withdrawal	19	41

CHAPTER 5. TRAINING PROCEDURES
  Section I. Garrison training
             General............................ 20   45
             References......................... 21   45
         II. Preparation of field exercises
             Purpose............................ 22   45
             Troop orientation.................. 23   46
             Control personnel orientation...... 24   47
             Evacuation plan.................... 25   48
             Terrain preparation................ 26   48
             Friendly representation............ 27   54
             Enemy representation............... 28   55
             Observer-instructor................ 29   56
             Principal instructor............... 30   58
             Aggressor control officer.......... 31   59
             Safety personnel (medics, roadguards,
               ambulance driver)................ 32   60
             Supply personnel................... 33   60
             Transportation..................... 34   61
             Communication...................... 35   61
             Critique........................... 36   61
             Critique checklist................. 37   62
        III. Grading
             Suggested system................... 38   64
             Grading sheet...................... 39   67
APPENDIX I. REFERENCES................................. 69
        II. STREAM CROSSING AND SMALL BOAT
              SAFETY PROCEDURES........................ 71
       III. REVIEW TRAINING.............................. 76
        IV. FIELD TRAINING EXERCISES................... 108
         V. PATROL TIPS................................ 204
        VI. INSTRUCTOR'S GUIDE (RANGER HISTORY).. 213

CHAPTER 1
INTRODUCTION

1. Purpose and Scope

a. This manual is a guide for field unit commanders in establishing and conducting a program of instruction in ranger type training within their units. It includes general information about ranger type operations, including those operations supported by Army aviation, and sets forth current concepts of ranger training.

b. Included in this manual is an outline of a 5-week ranger training program. This outline presents the program of instruction for two weeks of preparatory review training and ten field exercises for the remaining three weeks. It discusses methods of preparing the field exercises, critiquing, and grading of personnel.

2. Concepts of Ranger Training

a. Ranger training is *realistic, rough,* and to some degree, *hazardous.* It consists of a minimum of academic instruction. Training is designed to develop the individual's self-confidence, leadership, ability to command, and skill in the application of basic infantry techniques. Every effort is made to simulate actual combat conditions with a minimum number of restrictions on the play of the problem. The actual problems present logical missions that a small infantry unit would be called upon to perform. The terrain over which the problem is run includes, whenever possible, natural obstacles such as streams, lakes, swamps, steep mountains, and thickly vegetated areas. Conquering these obstacles adds greatly to the overall ability and self-confidence of the individual. The student is required to call on his mental stamina as well as physical endurance to complete the more vigorous problems. He is assigned a "buddy," to operate with throughout all training.

b. Training is not postponed because of inclement weather. The ability to operate in all types of weather may often give a military advantage to ranger trained personnel.

CHAPTER 2
RANGER TYPE OPERATIONS

Section I. GENERAL

3. Definition

Ranger type operations are overt operations conducted in enemy-held territory. Their duration depends upon the location of the target, terrain, type of mission, enemy capabilities, and the tactical situation. These missions include combat raids, combat patrols, ambush patrols, reconnaissance patrols, and any special type missions requiring highly skilled personnel. Normally, they will not require parachutist qualification.

4. Requisites for Success

a. Success in a ranger type operation requires skill in one or any number of the following: scouting and patrolling, infiltration, evasion, small boats, rock climbing, mountain expedients, cliff ascending, rappelling, fieldcraft, survival, military demolitions, knot tying, map and aerial photography reading, radio procedure, employment of weapons, and night firing.

b. Ranger trained personnel are capable of operating in all types of weather over seemingly impassable terrain containing natural and manmade obstacles. They are capable of capitalizing on speed, surprise, and stealth to accomplish their assigned mission.

Section II. AIRCRAFT SUPPORT OF RANGER TYPE OPERATIONS

5. Tactical Use

a. The maximum use of all available support is considered in the planning phase of every ranger type operation. A major type of support that is always considered is Army tactical transport aircraft (fixed-wing and helicopter). Often the very nature of a ranger mission dictates the use of Army aircraft.

b. Due to their unique capabilities, Army tactical transport aircraft can be used in practically any kind of tactical situation; for example, movement to contact, offensive, defensive and retrograde operations, and special operations such as missions behind enemy lines. The terrain and extreme weather conditions which characterize mountain, desert, jungle, and Arctic operations retard movement and place a great deal of strain on equipment and men. The use of tactical transport aircraft, particularly helicopters, facilitates movement and reduces the time that equipment and troops are exposed to these stresses. Included within the capabilities of the aircraft are reconnaissance, security, supply and resupply, evacuation of wounded, and the capability to move troops from point to point practically any place in the battle zone.

c. The decision to use tactical transport aircraft depends largely on the type of operation, the nature of the terrain in the area, and the situation. Once the decision is made, the aviation unit supporting the operation is placed under the operational control of the ranger patrol leader.

d. If the operation is so large and complex as to warrant personnel being used for the purpose of aircraft control only, pathfinders should be requested from a higher headquarters. These pathfinders are specially trained Army personnel whose primary mission is to aid in the navigation and control of tactical transport aircraft in the objective area. The inherently small size ranger operation does not normally require pathfinder support.

e. See TM 57-210 for the characteristics, concepts, missions, capabilities, and limitations of Army tactical transport aircraft.

6. Operations Against Infiltration and Guerilla Action

a. Air landed forces are particularly suited to operations against enemy infiltrators and guerillas. In daytime, reconnaissance aircraft are employed to locate infiltrators; air landed patrols then follow up by investigating suspicious localities and destroying or capturing any enemy infiltrators. During periods of limited visibility, tactical transport aircraft position and support outposts and patrols, especially in difficult terrain likely to be used by infiltrators.

b. Small numbers of air landed troops can patrol extensive areas, and centrally located reserves can surprise guerilla bands in their hideouts; or they can be employed rapidly to reinforce installations and columns under attack. Guerilla tactics of blocking routes of reinforcement when attacking installations or ambushing columns are readily combatted by air transported reinforcements. Air landed forces exploit their mobility to attack guerilla bases of operations that are usually located in mountains, jungles, swamps, or other difficult terrain.

7. Night Operations

a. Advantages. Air landed forces may be employed effectively at night. They are less vulnerable to enemy ground and air fires, and the enemy has greater difficulty in determining the location of the main landing than in daylight operations. Small air transported units landing simultaneously at widely separated points may block movement, disrupt communications, and create general confusion while other ground or air landed operations are conducted.

b. Disadvantages and Problems. Night operations present certain disadvantages and special problems in comparison to daylight operations.

    (1) Both ground units and tactical transport aviation units require a high degree of training.

    (2) In selecting landing zones, greater stress is given to characteristics that assist landing than to placing units on or adjacent to objectives.

    (3) Ground units normally assemble after landing before proceeding on their missions, so assembly aids may be necessary.

    (4) Pathfinders at landing zones and sites and special aids to navigation are more necessary for movement and landing than in daytime.

8. Raid

The planning for a raid is similar to that for other tactical missions. The loading plan should provide for the transportation of prisoners and captured material. If the tactical transport aircraft are to be used for the withdrawal, this is planned for. The aircraft may remain in the objective area to facilitate transportation during the raid or to wait for the withdrawal. The decision to have the aircraft remain in the objective area is based on available concealment, the duration of the operation, enemy air capabilities, and the radius of action for the aircraft (figuring full loads for delivery and return). The withdrawal loading site(s) may be close to the objective(s) because the security units may withdraw on foot toward the objective(s) after the assault units have accomplished their mission. Or, the raiding force may break up into small groups to rendezvous with the aircraft at a predesignated point some distance from the objective(s).

9. Patrolling

The considerations involving the use of tactical transport aircraft with reconnaissance and security forces are applicable to patrolling. For reconnaissance patrols deep behind enemy lines, the following additional factors must be considered:

a. High performance reconnaissance aircraft may be used to gain information of the enemy and terrain in the vicinity of the objective(s) for planning purposes.

b. A decision must be made as to whether the tactical transport aircraft are to remain in the objective area. It may be necessary to use the aircraft to move the patrol from point to point when the area to be reconnoitered is large and the patrol is small. This use must be weighed with the problem of concealing aircraft movement within the objective area during daylight and darkness, and with the problem of refueling.

c. When the patrol is to be left in the objective area, plans must be made for the aircraft to return to a designated place at a designated time.

d. When all or elements of the patrol are to be delivered in the objective area at night or during other periods of limited visibility, the pilots must be able to navigate, hover, and land without the aid of pathfinders.

e. When deep reconnaissance patrols are planned, secrecy should be insured by moving to the objective area during periods of limited visibility.

10. Operations in Special Areas

a. Mountain. Transport helicopters can place a security echelon on critical heights and at distances unattainable for troops moving on the ground. Helicopters, within altitude limitations, are valuable as prime movers for direct fire support weapons because of the ease in which these weapons can be moved to dominating terrain. (Army observation aircraft can be used to provide observation over wide areas to the front of friendly forces and to perform surveillance missions between friendly strong points.) "Dead spaces" in radio reception may be overcome by establishing aircraft retransmission stations, and helicopters may be used to lay wire over otherwise inaccessible terrain.

(1) Due to inadequate road nets, logistical support may be completely dependent upon Army tactical transport aircraft. Resupply and evacuation installation sites may be farther from the line of contact, and supplies can be placed closer to the location where they are needed.

(2) When formulating plans for operations in the mountains, the possibility of sudden weather changes is considered. Alternate plans are prepared and alternate positions for air-transported forces are selected in the event the first choice becomes unattainable.

(3) Large landing zones are rare because of rough terrain; therefore, if a large force is to occupy only a few terrain features, it may have to be shuttled into position. Factors to be considered in selecting landing sites are—

- (a) The direction of wind drafts, snow, and ice covered slopes (which require external cargo and troops to be unloaded while the helicopters hover).
- (b) Adequate space for the rotor blades (so they will not strike a mountain side).
- (c) The necessity for pathfinders (to mark routes and landing sites for safe movement and night landings).
- (d) Approach and return routes (which are selected to take full advantage of the defilade and concealment afforded by mountains).

(4) The actions of small, semi-independent units in seizing or defending heights which dominate lines of communications, or of fighting to seize or block passes and defiles on routes of communications, are of increased importance.

(5) Special clothing for personnel and equipment for aircraft must be specified and issued prior to the operation. Troops must be thoroughly familiar with the use and maintenance of these special items.

(6) The techniques of recovering aerial delivered supplies and the selection, preparation, and operation of landing sites, drop zones, and loading sites must be understood by the units.

b. Jungle.

(1) When formulating plans for a jungle operation, emphasis is placed on providing mobility through the employment of Army tactical transport aviation. The use of tactical transport aircraft in this type operation presents these considerations—

(a) The ability of reconnaissance and security forces to cover large areas of surveillance in transport aircraft permits the selection of more dispersed objectives.
   (b) Because of the dense vegetation in jungles, observed fire support of an air landed operation from fire support units behind friendly lines and from units within the objective area is limited.
   (c) The range of air landed operations should be short if a group linkup is to be made early.
   (d) As in mountains, suitable landing zones are few in number, and a shuttle system may have to be employed, utilizing tactical transport helicopters. Landing zones should be selected close to objectives to take advantage of the concealment afforded by the jungle and to reduce the distance necessary for movement on foot through the dense undergrowth.
   (e) Waterways provide a means of communications and are an aid to navigation; these factors should be considered in the selection of objectives.
   (f) Approach and return routes should take advantage of the concealment afforded by the jungle.
   (g) Fewer resupply and evacuation installations can be used in the forward areas when forces are supported by tactical transport helicopters.
  (2) Supply and evacuation may be completely dependent upon Army tactical transport aviation. Fewer resupply and evacuation

installations are required in the forward areas when forces are supported by tactical transport helicopters. Prescribed loads are determined and supplies are palletized when necessary to facilitate loading and unloading.
   (3) These techniques should be considered for jungle use—
      (a) Troops can descend from hovering helicopters by ropes, rope ladders, or by parachute.
      (b) Loads can be released while the helicopters are hovering over a drop site.
      (c) Smoke or panels in trees can be used to mark landing zones and landing or drop sites.
c. *Arctic.*
   (1) Enemy installations and critical terrain features dominating enemy routes of communication and supply are appropriate objectives for air landed forces in the Arctic. The use of Army tactical transport aircraft presents these considerations—
      (a) Air landed advance and flank guards avoid much of the fatigue caused by foot movement in snow.
      (b) Aircraft enable reconnaissance and security forces to survey wide areas during the short span of daylight.
      (c) Air landed task organizations are small and compact for Arctic operations, and tactical transport aircraft can place fire support weapons closer to or within the objective area to avoid the difficulties of overland movement.

(d) Strong winds and blowing snow may interfere with or prevent the use of aircraft, so alternate plans for operating without them should be made.

(e) The brief period of daylight and the Aurora Borealis, (northern lights) which interferes with or prevents radio communication, should be considered in timing the operation.

(f) Suitable landing zones are plentiful except in the mountains.

(g) Approach and return routes should take advantage of the defilade and concealment of any rough terrain in the area.

(h) Operations logistically supported by tactical transport aircraft can be executed at far greater ranges than those supported by ground means, and fewer intermediate installations that are vulnerable to enemy raiding forces are needed.

(2) These techniques are appropriate for Arctic operations—

(a) Landing sites can be marked so that they will be recognized when terrain features are blanketed or their appearance is changed by heavy snowfall.

(b) Navigational aids must be planned to overcome the effect of "white-out" (loss of reference due to the skyline merging with the snow covered terrain).

(c) Shelters must be provided for maintenance personnel.

(d) Portable homing devices should be provided because some areas are not mapped, and recognizable check points are few.

(e) Army tactical transport aircraft must be completely winterized, and preheating may be necessary.

(f) Pilots should be trained for ski-, wheeled- or float-type aircraft. During the summer months, floats permit utilization of the many lakes and streams as landing zones. During the winter, these same lakes and streams are ice-covered and may be utilized as landing zones by ski-equipped aircraft. During the breakup and freezeup periods, neither ski nor floats are usable on fixed-wing aircraft. Therefore, operations are restricted to wheeled aircraft utilizing available landing and take-off areas.

d. Desert.

(1) The highly mobile reconnaissance and security units necessary in desert operations can be provided by making them air transported. If satisfactory navigational aids are available, aircraft can be employed successfully at night—a factor in timing an operation. Other considerations when using transport aircraft in desert operations are—

(a) Unobstructed landing zones large enough for mass landings are plentiful.

(b) Mountainous areas may offer the only concealment for approach and return routes.

(c) Tactical transport aircraft can resupply and evacuate combat, reconnaissance, and security forces, making it possible to decrease the number of lines of communications.

(2) These techniques are appropriate for desert operations—

  (a) Oil or a similar substance can be used on landing and loading sites to minimize the operational difficulties caused by dust and sand.

  (b) Pathfinders can use navigational aids such as smoke, panels, and electronic devices on routes and landing zones to overcome the difficulty of terrain orientation.

CHAPTER 3
AMBUSH AND ROADBLOCK TECHNIQUES

11. Introduction

a. An ambush is a surprise attack from a hidden position against a moving enemy. It is the major operational method of attack for guerillas, but can be employed successfully by any infantry unit. The ambush can be used practically anywhere in the combat zone. Operations against guerillas and infiltrators in rear areas may utilize ambush techniques. Forward units may employ an ambush to kill or capture enemy patrols. Patrols may be sent into enemy territory with the mission of ambushing and destroying an enemy vehicular convoy train or killing or capturing troops of a foot column.

b. A roadblock is a valuable adjunct to the ambush. Often it is established as a part of an ambush; for example, an ambush patrol may be given a mission of destroying a vehicular convoy and establishing a roadblock with the wrecked vehicles. Some type of temporary roadblock is often employed to assist in stopping the enemy in an area selected for the ambush. This roadblock is defined as temporary because it differs from a well-constructed and defended roadblock that is employed to deny the enemy use of a route of movement. Examples of temporary roadblocks are felled trees, ditches, land slides, and hasty minefields.

c. Whether the ambush utilizes a roadblock or not, the key to success against a foot or motor column is the element of *surprise*.

12. Conduct of an Ambush

The ambush is executed with surprise, shrewdness, and violent determination. Proper intelligence, planning, and coordination are necessary when formulating plans for a successful ambush operation. Some factors to be considered are—

a. Knowledge of Enemy Tactics and Habits. The commander with the responsibility of executing an ambush should have the maximum knowledge of enemy tactics and the manner in which the enemy will probably react to the ambush. All available intelligence sources should be tapped to assist in providing usable information. Care must be exercised to provide defense against both limited and determined countermeasures by the enemy. Also, the availability and location of reserve enemy units positioned to reinforce the unit to be ambushed must be considered; and plans must be made to isolate the ambush site to prevent this reinforcement.

b. Selection of the Ambush Site. When selecting the ambush site, a careful study must be made utilizing maps, aerial photos, and personal reconnaissance, where possible. Consideration must be given to the availability of natural obstacles that may be used. A good example of this is a route with a steep cliff or swamp on one flank and good positions from which to deliver a heavy volume of fire by the ambusher on the other flank (fig. 1). The absence of these natural obstacles may be overcome by substituting minefields or road craters. The most successful ambush site is one in which the enemy is cut off and unable to deploy his forces. Tricks or ruses to lure him into the ambush sector should not be overlooked. For example, road signs can be placed to confuse enemy columns; feints or limited objective attacks can be made at one site in order to direct the unit to be ambushed to the desired location. Commanding ground, existing cover, and concealment should be fully utilized.

c. Routes to and From the Ambush Site. The route to and from the ambush site is carefully selected to allow for secrecy in occupying the positions and speed and security in withdrawing from it. The maximum amount of cover and concealment is desirable. After entry into the ambush area, the route followed by the ambushing force is carefully inspected to remove all evidence of the force's presence. In planning the withdrawal, alternate routes are selected to avoid the possibility of enemy forces blocking the withdrawal of the ambush force. These routes are patrolled periodically to insure that enemy infiltrators do not occupy them.

d. Special Equipment Needed. The type of ambush dictates the need for special equipment. An ambush of personnel normally requires a large percentage of automatic weapons. Fragmentation and white phosphorous grenades, antipersonnel mines, rocket launchers or rifle grenades, demolitions, and incendiary devices can be used to immobilize and destroy vehicles and material. Ample communication equipment is necessary, including extra radios, telephones, pyrotechnics, and infrared devices.

e. Communications and Control. Control is necessary during the movement to, occupation of, and withdrawal from the ambush site. The most crucial time of the ambush operation is the moment the enemy arrives at the site. Communications must provide for the issuance of orders to open fire. If plans do not provide such a signal, the time of arrival of the lead element of the enemy at a certain location is designated as the time to open fire. A similar plan is provided for both the assault, if it is to be conducted, and withdrawal. Communication with local security elements and higher headquarters is desirable. Exacting control must be exercised to insure that the ambushing force is alert and silent. Once the force is in position, movement must be kept to a minimum. Care is taken to insure that the ambush is not executed prematurely. Assembly points are designated to assist in control during the withdrawal.

f. Rehearsal of Participating Troops. Rehearsals for the ambush operation are conducted on terrain similar to that which is to be used on the actual mission, and must include all personnel scheduled to participate in the action. Subordinate leaders, weapons crews, and security elements are briefed until all personnel know the exact sequence of events of the ambush and understand their duties thoroughly. Equipment is checked at the rehearsal and all weapons test fired, if necessary. All men should know the routes of withdrawal and location of assembly points (fig. 1).

g. Camouflage Measures. In no other operation is camouflage discipline more important than in the ambush. Personnel and weapons must blend with the surrounding area as much as possible, and all residue resulting from preparation of the site must be removed. Once the position is camouflaged, personnel must not move unnecessarily. Patience and staying power is a must to avoid premature disclosure of the ambush.

h. Coordinating Fire Plan. The fires of all weapons, whether long-range artillery or close-in automatic rifles, are tied into the fire plan. The time to open and cease fire, the assignment of sectors of fire, and the location of friendly forces are considered. Plans are made for isolating the ambush area to prevent the escape of the enemy and deny his reinforcement. The effectiveness of the ambush relies upon the *surprise* delivery of a large volume of fire. The fire should come from at least two directions and converge on the target. Care is taken to prevent friendly troops from firing into other positions of the ambush area when converging fire is used. To achieve surprise, fires of the ambush force do not commence except upon a prearranged signal. The withdrawal is considered in the fire plan to prevent the possibility of pursuit by the ambushed enemy or nearby enemy units.

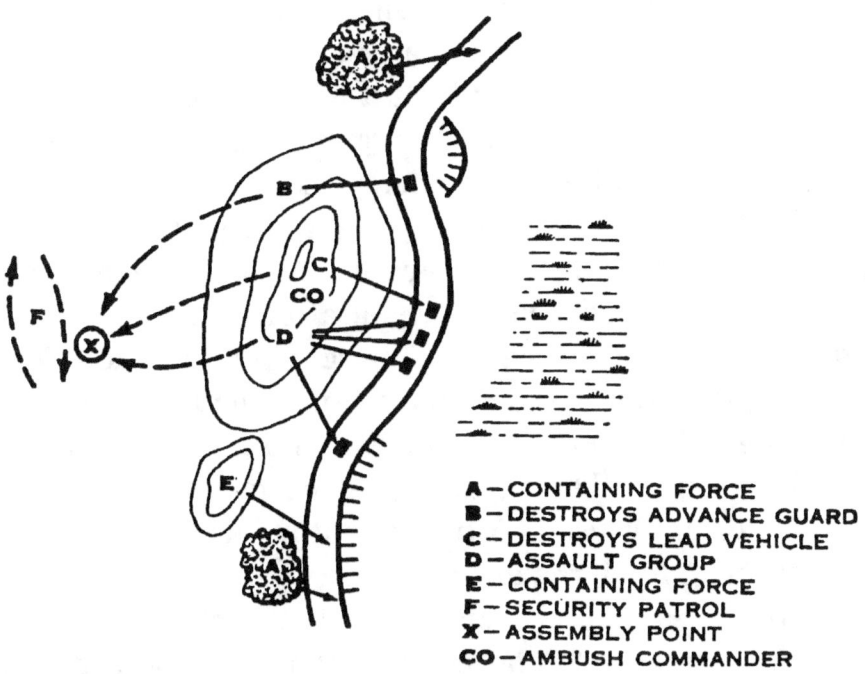

A — CONTAINING FORCE
B — DESTROYS ADVANCE GUARD
C — DESTROYS LEAD VEHICLE
D — ASSAULT GROUP
E — CONTAINING FORCE
F — SECURITY PATROL
X — ASSEMBLY POINT
CO — AMBUSH COMMANDER

Figure 1. Ambush position.

i. Use of Assault Groups. The use of "killer teams" or assault groups in conjunction with the ambush is often desirable. The purpose of this group is to physically move through the ambush site and assure the destruction of vehicles and material, search enemy dead, or any other duties considered necessary by the ambush commander. The action of this group must be planned, rehearsed in detail, and violently executed. It is usually of short duration. The nature of the ambush usually dictates the strength and employment of this group.

13. Establishing a Roadblock

a. The roadblocks discussed in this section may be either temporary (hasty) or deliberate. The temporary roadblock commonly used with an ambush is designed to halt the lead elements of the enemy force when it arrives in the ambush area and to "trigger" the ambush action. An example of a hasty roadblock is a tree wrapped with enough charges to fell it across an approach route when the enemy comes within the ambush site. Another example is an antitank mine buried in the route and set to explode when the enemy appears. The preliminary preparations for a hasty roadblock, such as placing charges or positioning vehicles, is completed as rapidly and silently as possible to keep the location, strength, and mission of the ambush secure. The deliberate (permanent) roadblock is normally used to block routes into our battle positions or to force the enemy into a position favorable to our defenses. It is included in the barrier plan of the overall defense. The construction of a deliberate roadblock depends on the time, material, and personnel available to build it.

b. The most common permanent roadblocks are—

(1) *Antitank ditches.* This roadblock is usually employed in conjunction with minefields and wire entanglements. It may be dug by hand tools or with bulldozers.

(2) *Side hill cuts.* This type roadblock is placed with demolitions by the employing units. When locating this roadblock the employing unit must consider the availability of bypass routes the enemy may use.

(3) *Road craters.* A road crater is a large hole in the center of a route of approach that is designed to deny the enemy the use of the route.

(4) *Log obstacles.* Log obstacles are either rectangular or triangular in shape and are usually filled with rocks or dirt. Log posts may be dug into the ground to form a similar type obstacle.

(5) *Abatis.* An abatis is constructed by felling trees at an angle of about 45° to the enemy approach. The trees are left attached to the stumps to prevent rapid removal.

(6) *Steel and concrete obstacles.* Roadblocks of steel and concrete may be constructed when personnel, time, and material are available. Dragonteeth, I-beams, ramps, or other designs may be used. For complete details of roadblock and obstacle construction see FM 5-15 and TM 5-310.

c. To be effective, the roadblock is placed in a position to deny the enemy the opportunity to bypass it or to sufficiently deploy his forces so as to conduct a strong attack against it. It is covered by accurate

fire. Full use is made of natural obstacles such as swamps, dense woods, or extremely rough terrain. Antitank and antipersonnel mines may also be used. A good roadblock should be—
  (1) Placed along a likely avenue of approach.
  (2) Strongly constructed.
  (3) Difficult to destroy.
  (4) Constructed with materials available locally.
  (5) Covered by accurate fire.
  (6) Concealed from long-range enemy observation.

14. Defense Against Ambush

a. In planning for defense against ambush, the planner must initially consider the friendly force available. A large vehicular column reinforced with armor reacts differently against attack than a small unit of foot troops without reinforcements. The small unit commander responsible for moving a unit independently through areas where ambush is likely must plan for—
  (1) The formation to be used.
  (2) March security.
  (3) Communication and control.
  (4) Special equipment.
  (5) The action to be taken if ambushed.
  (6) The reorganization.

  (a) *Formation.* A dismounted unit normally uses a formation that provides for all-round security while en route. Responsibility for this is assigned to subordinate commanders and their units. The distance between units in march depends on the terrain, visibility, and situation. March interval is determined by the amount of control

available and the ability of units to support each other in the event of an ambush. The interval is also great enough to allow each succeeding element to deploy when contact with the enemy is made. However, the distances are not so great as to prevent each element from rapidly assisting the element in front of it. The column commander should be located well forward in the formation but is not restricted from moving throughout the formation as the situation demands. Units are placed in the formation so they may distribute their firepower evenly throughout the formation. If troops are to be motorized, unit integrity is maintained when possible.

(b) *March security.* Regardless of whether the unit is on foot or motorized, security to the front, rear, and flanks is necessary when ambush is likely. A frontal security element is placed well forward with adequate radio or pyrotechnical communication with the main body. The security element is strong enough to sustain itself until followup units can be deployed to assist in reducing the ambush. However, the enemy may allow a security element to pass unmolested in order to attack the main body. If this occurs, the security element may be used to attack the ambush position from the flanks or rear in conjunction with the main action to destroy it. Flank security elements are placed out on terrain features adjacent to the route of march. They move forward either by alternate or successive bounds, if the terrain permits. This is often difficult because of the ruggedness of the terrain and the lack of transportation or communications. The next best thing is moving adjacent to the column along routes paralleling the direction of march. Rear security is handled similarly to frontal security, and plans can be made for the rear guard to assist in reducing the ambush either by envelopment or by furnishing supporting fire. Reconnaissance by fire of likely locations for ambush may greatly assist the security forces; however, extreme care must be exercised by the convoy commander when authorizing this expedient because of the possibility of unknown friendly units in the area. Light aircraft flying above the column on reconnaissance and surveillance missions increase column security. When available, fighter aircraft can provide column cover. Air to ground communication between these elements is highly desirable.

(c) *Communication and control.* All available means of communication are used, consistent with security, to assist in maintaining control of a small unit during movement when ambush is likely. March objectives and phase lines may be used to assist the commander in controlling his unit. Communication with security elements is mandatory. Detailed prior planning and briefing and rehearsals for all units will assist in control if an ambush does occur. Alternate plans are made to prevent confusion and chaos. Higher headquarters is notified as soon as possible of the ambush so it may alert other units in the vicinity.

(d) *Special equipment.* It is often necessary to provide the unit with additional items of equipment and weapons, especially when it moves through areas where guerillas or other enemy forces are likely to be encountered. It is desirable that vehicles have an automatic weapon mounted and manned. Pioneer tools and mine detectors are used to reduce roadblocks and minefields. Demolition equipment is used to destroy obstacles encountered en route. Additional spare parts may be needed to make "on the spot" repair of disabled vehicles. Ample communications equipment is always necessary, including identification devices to friendly aircraft such as panel sets or smoke grenades.

(e) *Action to be taken if ambushed.* If the unit is ambushed, the most important counteraction is for all available personnel to return fire as rapidly and as heavily as possible. At least one automatic weapon is manned on all vehicles throughout the column. On trucks being used as personnel carriers, the tarp and bows are removed and the tailgate lowered to the horizontal. Troops riding in these trucks remain alert at all times and are trained to detruck immediately and to return fire.

Necessity may require that they dismount even before the trucks are fully halted. Assistant drivers are assigned to all vehicles in case the driver becomes a casualty. If an ambush occurs, drivers are instructed to halt their trucks on the road. They do not pull off onto the shoulders because the shoulders may be mined. If trucks are used as lead vehicles, their floors are reinforced with sandbags to reduce the effect of mines.

(f) *Method of attack.* If the strength of the ambushed unit is adequate and attacking the ambushers is consistent with the unit's mission, envelopment is usually the most desirable method of attack. During the planning phase of the move, a holding element and an attacking element are designated. Each element is briefed thoroughly on its actions and any alternate plans necessary to meet different situations. For example, a plan calling for the advance guard to be the holding force would not succeed if the enemy allowed this force to pass unmolested. If the strength of the ambushed unit prevents their attacking by envelopment, the plan normally should be to break out of the immediate area rapidly to minimize casualties. This is possible because plans for an ambush do not normally call for pursuit of the ambushed unit. If a unit is surprised by the enemy, it tries to overcome him by returning all available fire immediately. This also allows the ambushed unit to deploy and maneuver.

(g) *Reorganization.* The reorganization after an ambush involves the use of rallying points, and plans for local security and rapid movement by all units. Care is taken to minimize the possibility of the enemy pressing the attack during this period, and plans are made to continue the original mission. Clearing minefields and other obstacles, evacuating wounded, and disposing of disabled vehicles present special problems.

b. All plans and preparations are wasted if personnel fail to be alert and vigilant at all times. This applies especially to security elements. Consequently, duties such as flank guard are rotated often.

c. The principles discussed in this section can be applied with modification to all relatively small combat units.

CHAPTER 4
CLIFF ASSAULT TECHNIQUES

15. Purpose and Scope

a. This chapter outlines the techniques of assaulting a cliff obstacle as might be found during an amphibious or waterborne raid. These techniques may, however, be used by troops encountering an obstacle of this type regardless of its location or the nature of the operation.

b. This discussion of cliff assault techniques involves a patrol's landing on a hostile shore, its actions on the beach, assaulting and scaling of the cliff, actions on the cliffhead, and actions during the withdrawal down the cliff.

16. Introduction

a. The element of surprise is an essential part of the successful raid. In order to increase the opportunity of surprise at the objective area, the leader of an amphibious raid carefully considers his choice for a landing place. In most cases, the easiest landing can be made on a steeply shelved sandy beach. The patrol leader should, however, consider avoiding such a landing place because it will probably be defended.

b. If all members of a raiding patrol are prepared to swim and are able to climb a rock cliff, there exists a chance of getting ashore unopposed. This unopposed landing enhances the opportunity for surprise later at the objective area. Therefore, it is often preferable to accept the disadvantage of landing upon a beach with physical difficulties such as rock cliff obstacles, rather than a relatively "easy" beach approach with its probability of being well defended.

17. Special Equipment

a. In addition to the equipment necessary for the conduct of the raid itself, the patrol may need special equipment for overcoming the cliff obstacle on the beach. Metal scaling ladders, climbing ropes, toggle ropes, rope ladders, grappling hooks, and "bear claws" might be used. Rockets with grappling heads may be necessary to carry the ropes over the cliffhead. It may be necessary for the first climbers to have additional mountaineering equipment such as pitons and hemmers, snap links, and sling ropes.

b. Communication equipment such as sound powered telephones may be found helpful in control at the beach and cliff. Engineer tape may be used for control on the cliffhead.

18. Initial Landing

a. General. The landing should be in two waves. There are three parties in the first wave: the climbers who establish the scaling ropes or ladders, the patrol leader, and the beach security personnel.

b. First Wave. The first men ashore secure the boats at the beach while the others disembark. Next ashore are the climbers. For a company size raiding party, there normally should be a total of six senior climbers who initially establish the ropes. Three of these climbers are designated the No. 1 climbers and three the No. 2 climbers. Once on the beach they move directly to the base of the cliff. The No. 1 climbers then begin their ascent. The No. 2 climbers at the base of the cliff tend the No. 1's ropes (①, fig. 2).

  (1) Third ashore is the beach security. These men take up defensive positions on either flank of the landing area at the foot of the

cliff. One man is designated the beach control officer. With the help of a messenger and radiotelephone operator, he establishes the beach control team. The radiotelephone operator establishes the base telephone. The remainder of the first wave takes up firing positions while awaiting the climbers to establish the ropes. The boats withdraw as soon as they are cleared of the raiding party personnel.

Beach secured, No. 1 ropes established

Figure 2. Initial landing.

(2) Upon reaching the cliff top, the No. 1 climbers secure their ropes.
   (a) It may be necessary to use "bear claws" for this if no object is available around which the rope can be tied. Care should be taken that the tope is not obviously exposed on the cliffhead. This can be done by placing the "bear claws" under a small section of sod. A bayonet or entrenching tool can be used to remove the sod. The claws can then be placed under the sod and the spikes pushed into the ground. The rope should then be secured to the ring and the sod replaced (②, fig. 2). If the rope is to be looped so that it can be pulled down later, then a small stone or object should be placed under the sod to allow the rope to move freely (②, fig. 3).

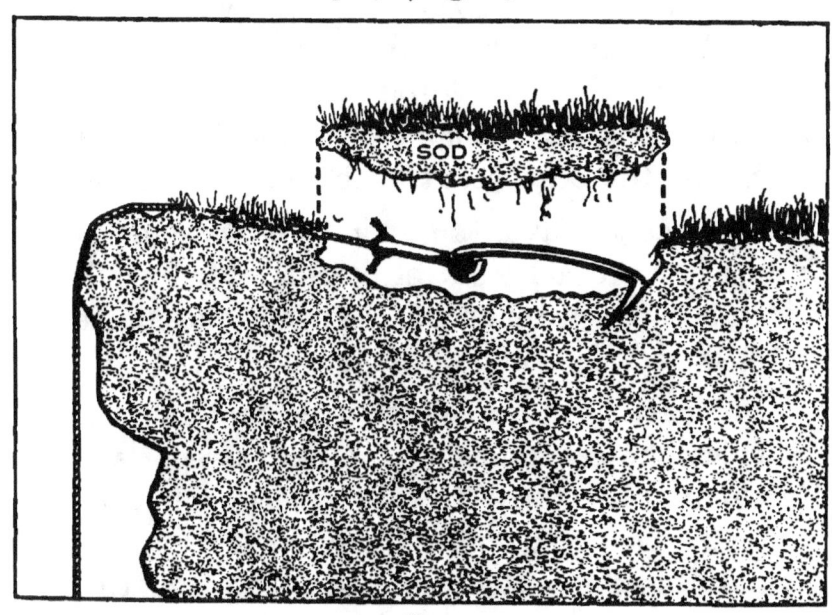

②
The rope and claws should not be exposed on the cliffhead
Figure 2—Continued.

(b) The No. 1 climbers then make a hasty reconnaissance of the cliffhead area to insure that the immediate area is unoccupied by the enemy. Having insured that the area is clear, they signal to the No. 2 climbers that they have cleared and secured the ropes. This signal should be two or three tugs on the rope. Each succeeding climber uses the same signal upon reaching the top to indicate the rope is clear. There should at no time be more than one man on any one rope. Upon being signalled, the No. 2 climbers immediately begin their climb. They carry a spare rope and "bear claw" with them (①, fig. 3). On reaching the cliffhead, they secure their spare ropes so that six ropes are available for climbing.

(3) As soon as the ropes are vacated by the No. 2 climbers, the patrol leader and security team rope up. They are followed as quickly as possible by the remainder of the first wave, leaving the beach control team below (fig. 4).

(4) The second wave, the main force, is signalled in by the beach control officer. Meanwhile, the security team on the cliffhead takes up defense positions on both flanks.

(5) The patrol leader selects a control point at which the remainder of the patrol reports on reaching the top. He also selects a location for his headquarters. White engineer tape is laid from the control point to the headquarters and from the control point to the two outer ropes in the climbing area (fig. 5). This

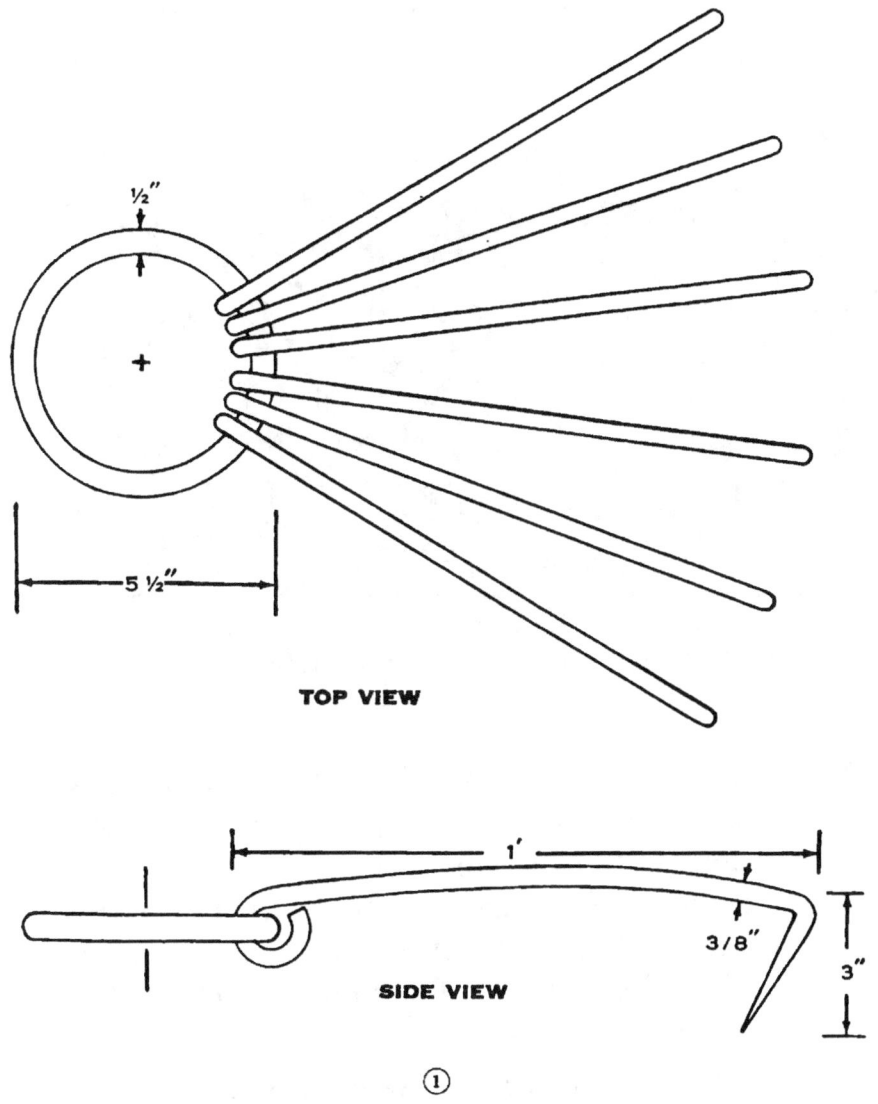

① Top and side views

Figure 3. Bear claws for securing ropes.

serves to canalize the remainder of the force into a central control point as they reach the top of the cliff. From this control point they are then directed to their various positions in the defense of the cliffhead.

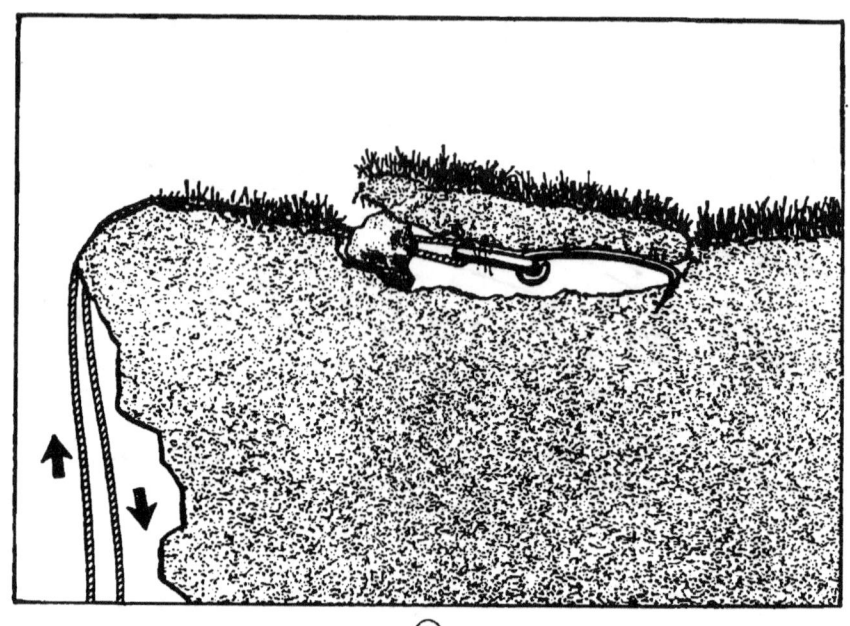

②

Ropes pull freely through the ring when sod is supported by a stone

Figure 3—Continued.

    (6) One man, designated the cliffhead officer, is located at the control point to direct the others as they move in.

    (7) The radiotelephone operator, with the patrol leader, establishes telephone communication with the operator in the beach control team at the base of the cliff.

c. Second Wave. The second wave initially takes up defensive positions on the beach and then moves for the ropes as directed by the beach control officer. The beach control officer insures that all ropes are in use.

    (1) The first men up are the subordinate leaders in the main force. Upon reaching the top, they move inland along the tapes to the control point, where they are directed to their re-

Figure 4. All ropes established—cliffhead secure.

spective sections in the cliffhead defense. They lay engineer tape as they move from the control point to their sections.

(2) The remainder of the force follows and takes up positions in their sections. As each man moves into the control point from the cliff edge, the cliffhead officer directs him to move along the tape leading to his section (fig. 6).

(3) As soon as the force is in position, a runner in each section reports to the cliffhead officer and takes in his tape on return to his section.

Figure 5. Control on the cliffhead.

The cliffhead officer then reports to the patrol leader that all men are in position.

(4) The patrol leader then leads the raiding party toward the objective, leaving the cliffhead officer and a detachment to defend the cliffhead.

(5) The cliffhead officer reorganizes the remainder of the force into a cliffhead team. The No. 1 and No. 2 climbers are left with this detachment. The No. 2 climbers initially take in the tapes from the ropes to the control point

and from the control point to the headquarters. The No. 1 climbers remain at their ropes. The radiotelephone operator in the beach control team then joins the cliffhead team and establishes radio contact with the patrol leader. One of the No. 1 climbers doubles his rope in preparation for the withdrawal.

Figure 6. Organization of the cliffhead.

(6) The cliffhead team then takes up security positions and awaits the return of the main force (fig. 7). The cliffhead team does not

commit itself to the enemy if at all possible. Enemy beach sentries are allowed to pass through the area unless there is a possibility of their seeing the main force. Every precaution is taken to prevent compromising the cliffhead area to the enemy. If sentries have to be eliminated, they should be taken silently. Care should be taken to properly conceal both the enemy dead and their equipment.

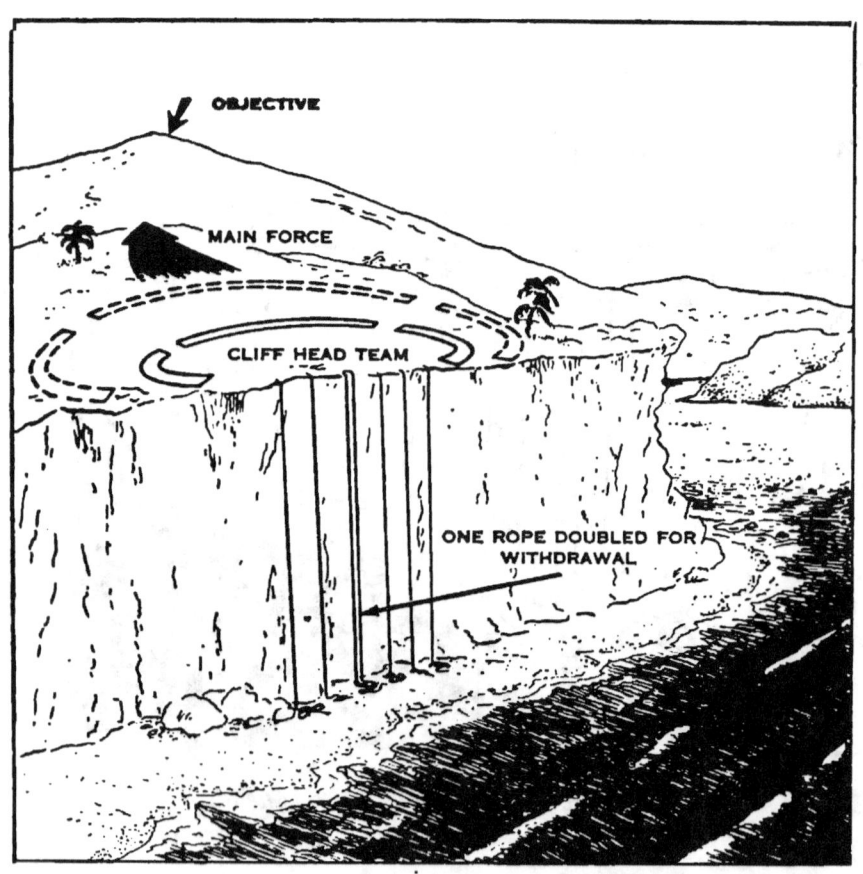

Figure 7. Reorganization of the cliffhead.

19. Withdrawal

a. If the situation permits, the main force returns to its original assembly area at the cliffhead, allowing for casualty evacuation and reorganization. Tapes from the control point and headquarters to the cliff edge may be relaid if necessary. In the event of close pursuit, the cliffhead team engages the enemy while the main force evacuates.

Figure 8. No. 2 climbers release their ropes and descend the No. 1 ropes.

No. 1 climbers release their single ropes and descend the remaining double rope

Figure 9. Evacuating cliffhead.

b. The patrol leader orders the party to prepare for withdrawal. On this command, the beach control officer signals the craft to come in. The beach control officer with his radiotelephone operator establishes a control point at the foot of the cliff. The No. 2 climbers go to the rope tops. On order to withdraw, the subordinate leaders of the main force withdraw to their ropes and send their machineguns to join the

Remaining No. 1 climber descends the double rope and then pulls it down

Figure 9—Continued.

cliffhead team. The cliffhead officer insures that all ropes are used. The subordinate leaders are the last of their sections to rope down. They report to the beach control officer on arrival. The cliffhead team then filters down and withdraws, followed by the patrol leader. The patrol leader's radiotelephone operator takes in his telephone lines as he goes. The No. 2 climbers cast off their ropes, recover their "bear claws," and descend the No. 1 ropes (fig. 8). The No. 1

climbers release all single ropes and descend the remaining double rope (①, fig. 9). The last man down pulls the double rope down upon reaching the base of the cliff (②, fig. 9). The remainder of the force then embarks. Throughout the withdrawal, each man takes every possible precaution to prevent any of his equipment from being left either on the cliffhead or on the beach. The area should be completely void of signs or traces of the patrol's presence in that area. The patrol leader and all of his assistant leaders should take extra precautions in controlling this factor during the withdrawal. Regardless of whether the raid is a success in material damage to the enemy, it, in all certainty, will be a success against his morale. His enemy has appeared on his ground, struck a blow, and vanished completely without a trace.

CHAPTER 5
TRAINING PROCEDURES

Section I. GARRISON TRAINING

20. General

a. The program of instruction (POI) contained in appendix III involves two weeks of preparatory review training in subjects required to develop the unit for the field training that follows. Paragraphs 9 and 10 of this appendix should be mandatory subjects because they are applicable to all units regardless of their location. The subjects in paragraph 14 are samples of those a unit may require in order to meet the hazards of unusual terrain or climate.

b. This training can be conducted in garrison with only a minimum requirement for training areas.

21. References

References listed in appendixes I and III are basic and will provide guidance for lesson preparation in support of the scope as assigned to the class. Army Subject Schedules listed in appendix I should be used to a maximum.

Section II. PREPARATION OF FIELD EXERCISES

22. Purpose

a. This section furnishes guidance to field unit commanders in the preparation of ranger training field exercises. The information that follows is of a general nature and can be applied to any or all of the field exercises listed in appendix IV.

b. To assist in the preparation of field exercises, instructional personnel should refer to the appropriate Army Subject Schedules listed in appendix I.

23. Troop Orientation

a. Troop orientation is performed by the principal instructor to familiarize the participating troops or students with the general tactical situation, to instruct them in the purpose of the problem, and to orient them concerning pertinent safety and administrative details. During this orientation, the troops are initially subjected to the realism of the training in which they are about to participate. From this point on, realism is essential in keeping with the general concepts of ranger training. Therefore, it is necessary that the orientation be presented in a realistic manner in order to inject enthusiasm and to stimulate the mental attitude of the participating troops.

b. The instructor presenting this orientation should be designated as a battle group S2 or S3, depending upon the type of mission outlined in the exercises. He should have those visual aids present which are normally found in combat under similar circumstances. A general and special situation constituting an operation order is presented during this orientation. The operations order may be given in its correct sequence, or the situation may be presented in a series of notes not necessarily in correct operations order sequence. The latter method is effective in training an individual leader in the proper method of rearranging his notes into sequence for presentation as his patrol order.

c. After the general and special situation, any one or all of the patrol members may be required to prepare a warning order and patrol order in writing. The observer-instructor then selects the particular individual who will actually issue the order. He may collect all of the written orders and check them for accuracy. This technique is beneficial in training all troops in the proper method of preparing a patrol order and does not restrict the training to the patrol leader. The time that the orientation should be presented is explained in each particular exercise.

d. The exercises should be tied in with each other in a logical, moving, overall situation. The continuity of the situation can be maintained by slight alterations in the wording of the general situations presented in each particular exercise.

24. Control Personnel Orientation

This orientation is conducted by the principal instructor to instruct the control personnel in their duties and conduct, to orient them concerning the play of the exercise, and to explain all pertinent safety and administrative details. It normally takes place at least one day prior to the participating troop's orientation. The control personnel consist of the following:

a. Aggressor control officer.

b. Assistant principal instructor.

c. Supply personnel.

d. Transportation personnel.

e. Medics.

f. Aggressor troops.

g. Friendly troops (does not include members of patrol).

h. Observer-instructor.

25. Evacuation Plan

Prior to the actual conduct of the exercise, the principal instructor orients all personnel in the location of the medical men, evacuation vehicles, and routes of evacuation. Aid men do not participate tactically in the exercise with the exception of being in the proper uniform, whether it be friendly or Aggressor. They do not disturb the tactical play of the exercise when rendering aid unless it is actually necessary. Actual injuries incurred by any participating troops are treated under the supervision of the patrol leader. Except in the most extreme cases, care is taken to maintain realism, and control personnel should avoid any administrative breaks or halts in the play of the exercise in order to treat injured troops.

26. Terrain Preparation

a. General. Each problem contains the suggested size of the area required for the problem and any special terrain features needed to support the purpose of the particular exercises. A map reconnaissance of the available training terrain establishes areas that appear suitable. A visual ground reconnaissance determines the best area for the exercise.

b. Friendly Areas. See figure 10. Having selected an area, the location of friendly installations and positions is decided. These positions should be as realistic as possible and in keeping with sound tactical doctrine. Each problem may contain any or all of the following friendly positions, areas, or installations:

    (1) Reserve area.

    (2) Detrucking point in rear of forward company.

    (3) Forward company command post.

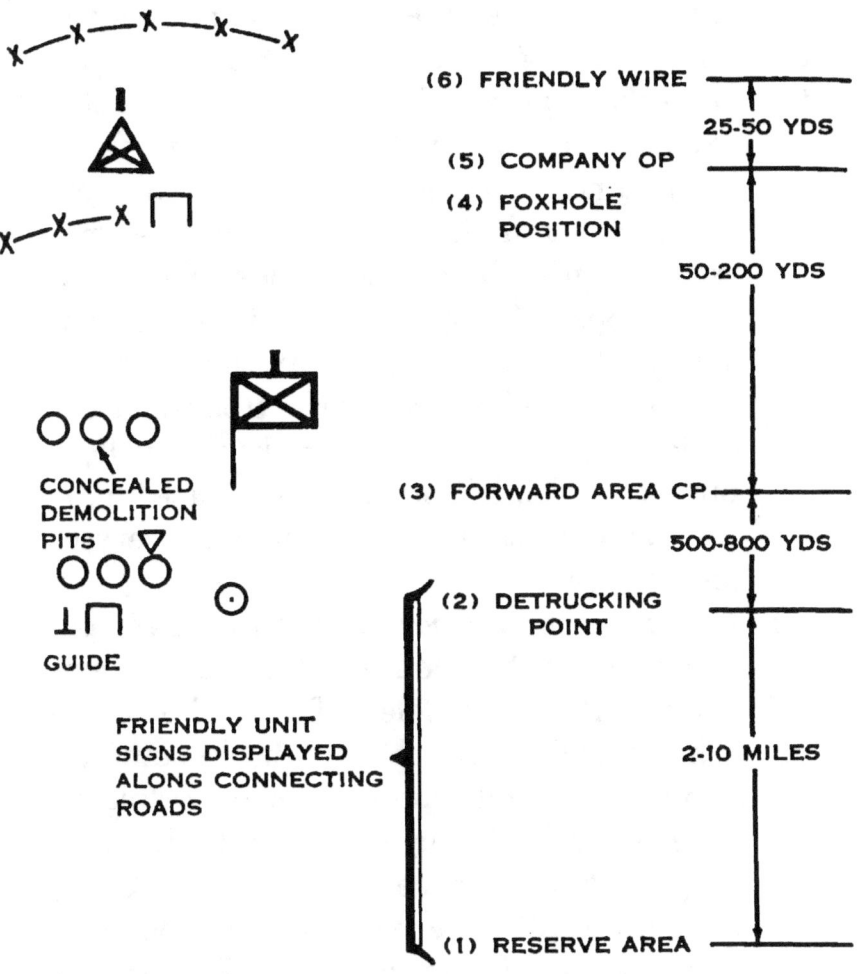

Figure 10. Friendly positions.

(4) Company forward battle area position.
(5) Company outposts.
   (a) *Reserve area.* The reserve area is located several miles behind the forward positions near a road net. It is concealed from Aggressor positions. Tentage, bunkers, foxholes, and various other types of material and equipment may be utilized in the con-

struction of the area so that it affords the appearance of a forward reserve area.

(b) *Detrucking point.* This point is concealed from enemy observation and allows for a hard stand for vehicles and a suitable area for turn arounds. There is a prepared position (foxhole) for the guide who meets the troops. Company unit signs are conspicuously placed in or near this point so they may be observed by the troops. Concealed personnel with simulator artillery bursts or demolition pits are placed near this point to simulate incoming artillery or mortar fire.

(c) *Forward company command post.* This position can either consist of a bunker emplacement or foxhole. There should be other visual aids such as radios, communication wire, and ammunition boxes placed conspicuously in or near the position to suggest that it is a command post. Again, demolition pits or concealed personnel with artillery simulators are placed near the position to simulate enemy, incoming artillery, or mortar fire. The position should be tactically sound. It is occupied by the forward company commander.

(d) *Company forward battle area position.* This consists of open foxholes or bunkers, barbed wire, or any other materials normally found in and around such a position. All positions are not necessarily manned.

(e) *Company outposts.* This position, a foxhole or bunker, is manned by one or two men to give the appearance of a listening post or outpost. Barbed wire is placed in front of the position as would normally be done, and a gap exists through which the patrol can depart.

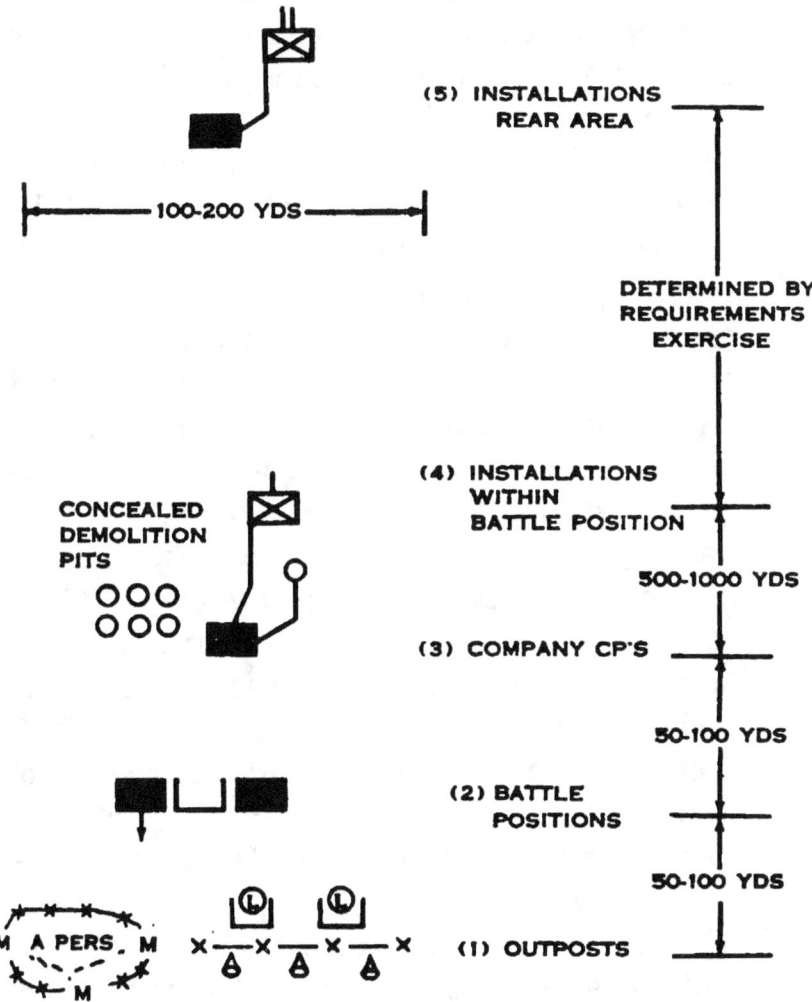

Figure 11. Enemy positions.

c. Aggressor Areas. See figure 11. The Aggressor positions may consist of any or all of the following:
  (1) *Outpost.* The Aggressor outpost can be constructed in the same manner as the friendly outpost. Roving patrols in front of the Aggressor positions may also be employed in those areas of likely patrol entrances not covered by actual emplacements. Dummy marked minefields and barbed wire can be improvised and used to canalize the patrols into occupied positions.
  (2) *Main battle position.* These positions, like all other positions and installations, should be located in areas that are tactically sound. Basically, they should be built and occupied to allow for flexibility of movement of available Aggressor troops. These troops should be able to move about the position, giving it the appearance of a well-manned battle position without making it obvious to the participating troops that there is only a representative group being employed. There should be a lateral road net behind the area selected for the battle position if possible. This allows for rapid movement of troops and facilitates control by the Aggressor control officer. Barbed wire should be laid and simulator booby traps, trip flares, and noise makers employed.
  (3) *Forward company command posts.* These positions can serve as control and supply points for the Aggressor. Concealed demolition pits should be near the position to simulate friendly artillery fire. Entrance into the

position should be easily gained from the rear to facilitate movement of supplies and personnel.

(4) *Installation within the main battle position.* A variety of installations may be constructed within the position to serve as objective sites for the participating troops. In addition to these installations, there should be Aggressor unit signs placed conspicuously along roads and trails to indicate Aggressor activity. Deserted buildings and installations may be available for use within the training areas. Examples of these areas are—

(a) Mortar positions.

(b) Forward regimental command posts.

(c) Artillery installations.

(d) Control and check points (traffic and/or security).

(5) *Rear area installations.* These can be treated in much the same manner as those in b, above. Examples of these are—

(a) Regimental and division command posts.

(b) Heavy artillery positions.

(c) Guided missile positions.

(d) Radar installations.

(e) Motor pools or supply depots.

(f) Critical installations including dams, bridges, and power plants.

d. *Other Areas.* This refers to the area between the friendly and Aggressor battle position. The terrain here should remain as much as possible in its natural state, thus providing natural obstacles and barriers to the patrol. If possible, there is no activity in this area other than that called for specifically in a particular exercise. This area represents no-man's land and is void of any activity or installations that detract from the realism of the training.

27. Friendly Representation

a. Number. As many of the prepared positions in the forward company area as possible are occupied by friendly troops. In addition to personnel called for specifically in a particular problem, the following minimum will, in most cases, be necessary for the occupation of forward positions of one lane (one patrol):

(1) One guide.
(2) One outpost man.
(3) One friendly forward company commander.

b. Uniform. Troops are dressed in the proper United States Army uniform and should give the appearance of having been in combat for some time.

c. Equipment. The individual equipment of the friendly force includes that normally found on a person in a similar capacity during actual combat.

d. Duties and Conduct. The guide's duties consist of meeting the patrol at the detrucking point and leading it to the forward company post to the outpost.

(1) The forward company commander is prepared to discuss the tactical situation concerning his immediate front with the patrol leader. He is also able to answer all questions and offer aid concerning coordination with the patrol. How much information he volunteers and the amount of coordination he offers should be decided upon by the principal instructor prior to the actual operation of the exercise.

(2) The outpost man is generally aware of the tactical situation. He is prepared to coordinate with the patrol leader concerning the time and method of the patrol's return. He is acquainted with the method of challenging and the proper use of the password.

28. Enemy Representation

a. Number. The number of Aggressors is determined by the situations and requirements of the problem.

b. Uniform. Forward of the friendly positions all personnel, including the principal instructor, Aggressor control officer, inspectors, and administrative personnel are dressed in complete Aggressor uniform. This includes *all* personnel except members of the patrol or troops. It is imperative that this be strictly adhered to and controlled in order to maintain realism and prevent destruction of the tactical atmosphere of the exercise. Care is taken also to have all vehicles forward of friendly lines properly marked with Aggressor markings. In keeping with realism, all of the above personnel upon their return into friendly lines revert back to the proper U. S. Army uniform, doing so in such a manner as not to make it obvious to participating troops that they represent Aggressor forces. The same applies to Aggressor marked vehicles.

c. Equipment. The Aggressor's individual equipment includes those items normally found on a person in a similar capacity during combat.

d. Duties and Conduct.

(1) The employment of Aggressor troops depends upon the type exercise being presented. The

principal instructor is responsible for the Aggressor plan of action. The Aggressor control officer is responsible for the proper execution of this plan. This plan includes Aggressor action in the main battle position, Aggressor rear area, at patrol objective sites, Aggressor ambush sites forward of their main battle position, and at any other location where the Aggressor is being employed. This plan is presented to the Aggressor troops during their orientation.

(2) The action of the Aggressor must be logical and, as far as possible, tactically sound in nature. Aggressor tactics are used. In many cases, the enthusiasm of the participating troops depends on the way the Aggressor is tactically employed, his enthusiasm, and the way he plays the game.

29. Observer-Instructor

a. Number. At least one observer-instructor accompanies each patrol during the conduct of a particular exercise. When patrols larger than one squad are employed, it is desirable to use at least one observer-instructor with each squad.

b. Uniform. The observer-instructor's uniform is the same as that designated by the patrol leader in his patrol order.

c. Equipment. Other than equipment specifically called for in a particular exercise, the observer-instructor should have a map of the training area and a compass in order to know at all times the location of the patrol.

d. Duties and Conduct.

(1) During the actual conduct of an exercise, the observer-instructor becomes a member of the patrol. He is prepared to give an accurate and critical report on the conduct of the entire patrol at a critique following the exercise. The critique provides the observer-instructor with the opportunity to teach. During the actual patrol he offers no criticism or aid to the patrol leader, but he makes mental and written notes of the actions to prepare himself for the conduct of the critique. Only in the case of an extreme emergency, such as the possibility of loss of life or damage to government property, does he interfere with the patrol leader's decisions or actions.

(2) If desired by the principal instructor, the observer-instructor can, throughout the course of the exercise, declare casualties and change the command of the patrol. In doing so he takes care to maintain the realism and continuity of the exercise.

(a) In a case where there is no contact with the Aggressor, but the observer-instructor desires to change patrol leaders, he might have the patrol leader pretend to break a leg. The second in command or some other person designated by the observer-instructor then takes over the command and arranges for this man's disposition. After the new patrol leader's action concerning this man is completed, the observer-instructor instructs the casualty to join the other members of the patrol for the continuation of the problem. The man designated as a casualty simulates being one only so long as the new patrol leader is deciding on and supervising the man's disposition.

(b) When the patrol is in contact and engaged with the Aggressor, the observer-instructor may declare casualties as he so desires. This is the most advantageous time in which to change command because it tests the new patrol leader's initiative, decisiveness, and other leadership qualities under trying conditions. Where the patrol and Aggressor actions are stalemated, the observer-instructor assesses casualties on both sides in order to force positive action and insure continuation of the exercise.

30. Principal Instructor

a. Number. There is one principal instructor for each problem.

b. Uniform. The principal instructor's uniform depends upon his location. If he is within friendly lines, it is the proper U. S. Army uniform. If forward of friendly lines, he wears complete Aggressor uniform.

c. Equipment. No special individual equipment is required.

d. Duties and Conduct. The principal instructor is responsible for the orientation of all personnel involved in the exercise and for the preparation of the terrain. He instructs and supervises the Aggressor control officer, the observer-instructor, the friendly and Aggressor representative groups, and all control and administrative personnel. He is responsible for requesting and procuring the necessary communication equipment, ammunition, transportation, rations, and any other equipment necessary to support the exercise. He is also responsible for the proper maintenance of this equipment during the exercises and the proper turn-in of the equipment at the completion of the exercise. He provides for emergency evacuation and instructs all personnel in the operation of this plan. He is responsible for instructing all personnel in pertinent safety regulations and in requiring their adherence to these regulations. The principal instructor supervises the debriefing and critique of the participating troops.

31. Aggressor Control Officer

a. Number. One officer or senior NCO is designated Aggressor control officer.

b. Uniform. While forward of the friendly positions the Aggressor control officer wears proper Aggressor uniform.

c. Equipment. Although it is not absolutely necessary, it is desirous that the Aggressor control officer be armed and have a map of the training area.

d. Duties and Conduct. The Aggressor control officer is responsible for the proper execution of Aggressor actions as outlined by the principal instructor in the orientation. He assists the principal instructor by supervising the Aggressor troops (including administrative personnel within the positions) in preparing their position, maintaining their equipment, adhering to safety regulations, execution of the evacuation plan if necessary, and, in general, the Aggressor play of the problem.

32. Safety Personnel (Medics, Roadguards, Ambulance Driver)

a. Number. The number and types of safety personnel are designated in each particular exercise.

b. Uniform. Uniform, depending on the men's location, is either U. S. Army or Aggressor.

c. Equipment. Safety personnel will have the special equipment necessary to accomplish their assigned task. In the case of a medic, an aid bag and litters may be desired. Ambulances or aid vehicles within Aggressor position are marked as Aggressor vehicles.

d. Duties and Conduct. See paragraph 24 for conduct of aid men and ambulance drivers during exercise. Roadguards are placed tactically into the play of the problem, regardless of whether they are within friendly or Aggressor positions. Safety personnel maintain the realistic continuity of the problem as far as possible.

33. Supply Personnel

a. Number. One NCO is designated as supply NCO for the exercise.

b. Uniform. U. S. Army uniform.

c. Equipment. No special equipment is required.

d. Duties. The supply NCO assists the principal instructor in requesting, procuring, issuing, and receiving all supplies necessary to support the exercise. He insures that the equipment is serviceable prior to issue and that it is properly cleaned and serviced prior to turn-in. He reports all items lost or damaged. He insures that safety regulations are observed when issuing, receiving, and storing explosives and ammunition.

34. Transportation

a. Number. The vehicular requirement is based upon the number of vehicles necessary to support the problem. Availability of vehicles is a consideration.

b. Uniform. Uniform, depending on the men's location, is either U. S. Army or Aggressor.

c. Equipment. No special equipment is required.

d. Duties and Conduct. Vehicle drivers do not participate tactically in the exercise with the exception of being dressed in the proper uniform and having the vehicles marked properly, whether it be friendly or Aggressor.

35. Communication

Radios, wire, or other methods of communication are integrated into all the exercises for control and safety purposes. They are also employed in the tactical play of the exercise, if desired. A control net consisting of friendly personnel and Aggressor personnel, each with different call signs, can be utilized. A separate net may be established for the patrol. In the case of the patrol net, the principal instructor establishes a station representing friendly elements to answer the patrol's calls and to communicate with them.

36. Critique

a. Purpose. The critique provides an opportunity to correct errors. It also provides an opportunity to compare exercises.

b. Conduct. Make the critique constructive by—
  (1) Briefly reviewing the action.
  (2) Pointing out the troop's achievements during the exercise.

(3) Pointing out the major errors noted and giving suggestions.
(4) Encouraging the men to ask questions that will clarify their understanding.
(5) Summarizing the lessons learned.
(6) Creating in the troops a feeling of accomplishment and a desire for continued achievement in training.

c. Conferences. Generally, the critique can be given most effectively by conference because this method encourages a 2-way exchange of ideas and thoughts between the observer-instructor and the troops. Guard against antagonizing and discouraging the group. Do not present a long list of deficiencies; avoid strong criticism of an individual or a unit in the presence of the entire group. Sometimes it is advantageous to conduct several critiques—one for the unit and a separate critique for the patrol leaders. This avoids possible resentment or lowered morale of any men or units taking part in the exercise.

d. Offer Solutions. Keep in mind the purpose of the training exercise. Recognize men who make outstanding contributions to their team's performance, and call attention to any errors or incorrect tactics without becoming personal. When errors are noted, give the correct solutions. When more than one solution is possible, give a preferred solution. Emphasize that other solutions are permissible, provided the fundamental points are correct and sound principles are followed.

37. Critique Checklist

The following is a general critique checklist for use by the observer-instructor. The remarks are applicable to all field exercises outlined in this text, and they should be used as a guide during the conduct of the critiques. When conducting a critique, it is recommended that the following list be used, together with the list contained in the exercise outline:

a. Was the warning clear, complete, and concise?

b. Were proper and adequate items of equipment and ammunition selected?

c. Did the patrol leader make a thorough map study prior to his reconnaissance?

d. Was the reconnaissance complete, including coordination with forward area personnel?

e. Was the coordination in the rear area complete?

f. Was the patrol order clear, complete, concise, and issued in a forceful, confident manner? Was it tactically sound?

g. Were visual aids employed during issuance of the patrol order?

h. Were patrol members properly prepared, inspected, and rehearsed prior to the patrol's moving out?

i. Did the patrol pass through friendly units in the proper manner?

j. Did the patrol listen for signs of enemy ambush just outside of friendly lines?

k. Was the formation suitable to terrain, cover, concealment, visibility, and proximity to known enemy positions?

l. Were the pace, point, and compass men properly utilized?

m. Was navigation accurate?

n. Was control maintained at all times?

o. Was security present and adequate at all times?

p. Were subordinates properly utilized?

q. Were signals used properly within the patrol?

r. Were rally points designated?

s. Were rally points easily distinguishable and tactically sound?

t. Were time elements and orders from higher headquarters adhered to? Was the mission accomplished?

u. Was the location of the patrol known to the patrol leader at all times?

v. Did the patrol leave any visible signs that would indicate its presence in the area?

w. Was *all* information reported in the debriefing or patrol report?

x. Did the patrol leader, his subordinate leaders, or any member of the patrol violate any of the "patrol tips"? See appendix V.

y. Did patrol leader display leadership traits such as knowledge, courage, initiative, decisiveness, tact, justice, dependability, bearing, endurance, enthusiasm, unselfishness, and integrity?

z. Did the patrol leader have the confidence, respect, obedience, and cooperation of all members of the patrol?

Section III. GRADING

38. Suggested System

a. The principal instructor may desire the observer-instructor to grade the performance of the various patrol leaders. The observer-instructor can do this along with his other duties, regardless of the number of leaders designated during the conduct of one of the exercises. The leader is graded on that phase in which he served in a command capacity. If one individual remains the leader throughout the entire exercise, only one grade is necessary. Two patrol leaders during the exercise of one patrol constitute the basis for two grades, etc. If more than one patrol is run during a single field exercise, several grades from each patrol may be obtained.

b. If the principal instructor desires to control the number of grades and the time an individual is to remain in a leadership capacity, he may assign a specific number of grades to be obtained. He does this by designating "phase lines," points where leaders can be logically changed. The observer-instructor can then be instructed to change the leaders at a given location (phase line) prior to the actual conduct of the exercise (fig. 12). To obtain four patrol leader grades during one patrol, the "phase lines" are assigned as follows:

(1) *Patrol leader No. 1* (phase I). Leads patrol initially and is relieved upon the patrol's arrival at the friendly forward positions.

(2) *Patrol leader No. 2* (phase II). Leads patrol through the friendly forward areas and to the objective.

(3) *Patrol leader No. 3* (phase III). Conducts the action at the objective.

(4) *Patrol leader No. 4* (phase IV). Conducts withdrawal from objective and leads patrol back to friendly positions.

c. When more than one patrol is being run in the same exercise, the observer-instructor with each patrol is required to get four grades. Figure 13 shows a method of obtaining six grades. In designating different patrol leaders by use of the "phase line" system, realism should be maintained. When a patrol leader is relieved, he should be declared a casualty by simulating explosion of a mine or artillery.

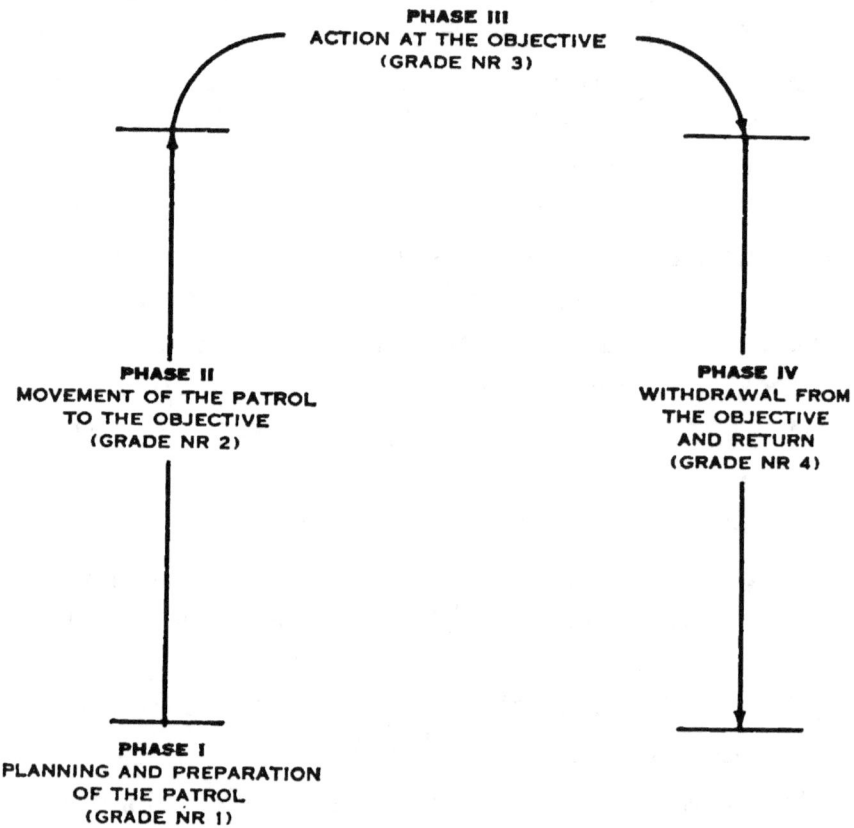

Figure 12. Phase I including grading for one patrol (four grades).

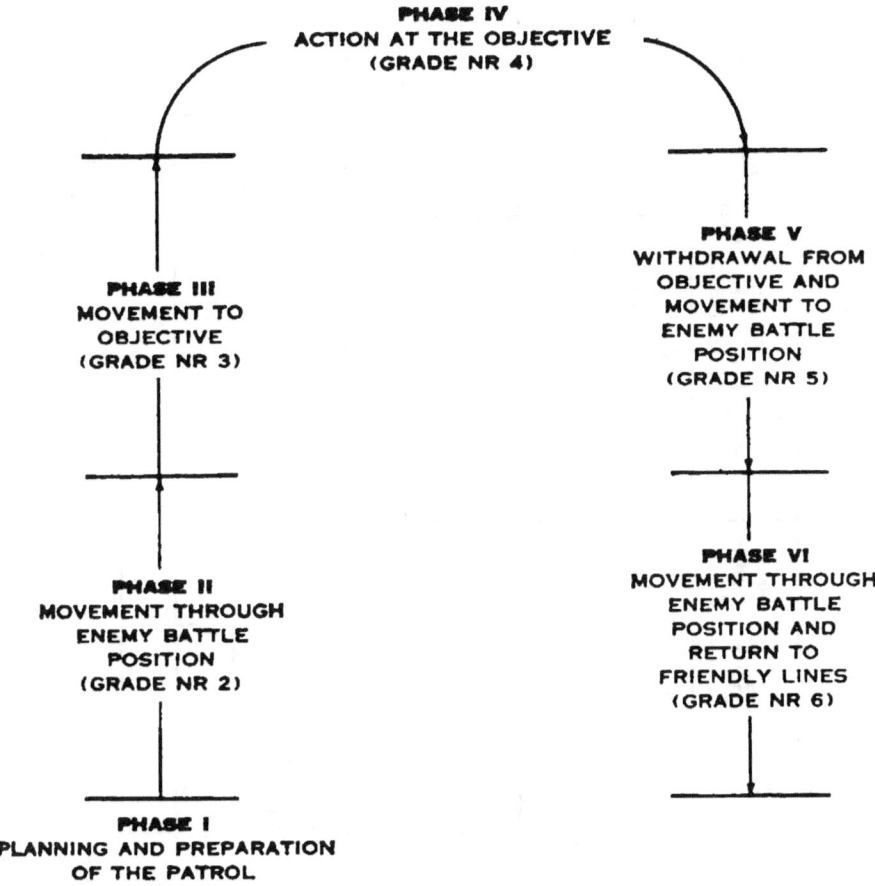

Figure 13. Phase line grading for one patrol (six grades).

39. Grading Sheet

The following is a recommended score sheet on which the observer-instructor prepares the grade. The reverse side of the sheet can be used for comments.

GRADING SHEET

Name of patrol leader _____

Report of leadership ability	Superior 10	Outstanding 9	Excellent 8	Very good 7	Good 6	Satisfactory 5	Poor 4	Very poor 3	Unsatisfactory 2	Undesirable 1	Value
1. Attitude											
2. Knowledge											
3. Forcefulness											
4. Control of Men											
5. Initiative											
6. Decisiveness											
7. Judgment											
8. Dependability											
9. Enthusiasm											
10. Endurance											

Problem Value

Signature of Observer-Instructor _____   Grade _____   SN _____

APPENDIX I
REFERENCES

FM 5-15	Field Fortifications.
FM 5-25	Explosives and Demolitions.
FM 7-10	Rifle Company, Infantry Regiment.
FM 7-40	Infantry Regiment.
FM 20-32	Employment of Land Mines.
FM 21-20	Physical Training.
FM 21-26	Map Reading.
FM 21-30	Military Symbols.
FM 21-75	Combat Training of the Individual Soldier and Patrolling.
FM 21-150	Hand-to-Hand Combat.
FM 30-15	Examination of Personnel and Documents.
FM 30-102	Handbook of Aggressor Military Forces.
FM 31-21	Guerilla Warfare.
[O]TM 5-310	Military Protective Construction.
TM 9-1900	Ammunition, General.
TM 57-210	Air Movement of Troops and Equipment.
DA PAM 21-46	Behind Enemy Lines.
ASubjScd 5-6	Explosives and Demolitions (Demolition of Equipment and Supplies).
ASubjScd 6-30	Umpiring and Aggressor Forces.
ASubjScd 7-2	Rifle Squad Tactical Training.
ASubjScd 7-9	Patrolling.
ASubjScd 7-20	Rifle Squad Tactical Exercises.
ASubjScd 7-30	Rifle Platoon Tactical Exercises.
ASubjScd 7-40	Rifle Company Tactical Exercises.
ASubjScd 7-53	Drop Zone Assembly, Day.

ASubjScd 7-54	Drop Zone Assembly, Night.
ASubjScd 21-1	Concealment and Camouflage.
ASubjScd 21-7	Intelligence Training.
ASubjScd 21-9	Maps, Compass, and Aerial Photograph Reading; Elementary Sketching.
ASubjScd 21-13	Signal Communications, Elementary
ASubjScd 21-16	Anti-Infiltration and Anti-Guerilla Warfare Training.
ASubjScd 21-20	Individual Day and Night Training.
ASubjScd 21-23	Land Mine Warfare (Mines and Booby Traps).
ASubjScd 21-25	Rifle Squad Tactical Training.
ASubjScd 21-26	Squad Patrolling.
ASubjScd 21-28	Bayonet.
ASubjScd 30-9	Combat Intelligence.
ASubjScd 30-15	Advanced Map and Photograph Reading.

APPENDIX II
STREAM CROSSING AND SMALL BOAT SAFETY PROCEDURES

1. Introduction

This appendix outlines recommended procedures that the principal instructor might use for control and safety during problems in which student patrols cross streams or rivers when bridges are not available. It also outlines safety procedures that might be followed in those problems in which small boats are used by student patrols.

2. Duties of the Principal Instructor

When a problem requires students to cross a river, stream, or body of water over five feet deep, the principal instructor—

a. Informs the observer-instructors of the hazards in the problem and reviews the procedures outlined below.

b. Reminds the observer-instructors of their responsibilities as safety officers.

c. Insures that supply issues the students enough one-half-inch or three-fourth-inch rope to breach the river or streams in the problem.

d. Insures that two power boats are on the river and within hearing distance of the students. One boat should be located upstream and one downstream from the crossing sites. In order to maintain realism, the boats should be concealed from the students. If possible, each boat should contain three men: boat operator, qualified swimmer, and a medic.

3. Duties of the Observer-Instructor

The observer-instructor who accompanies the patrol—

a. Checks the students before departing base camp to insure that they have in their possession the rope issued by supply and their individual survival ropes. Survival ropes issued to each student should be one-fourth-inch manila rope 8 to 12 feet in length.

b. Obtains the names of weak or nonswimmers. These students are observed and should not carry extra equipment such as machine guns or radios while actually crossing the rivers.

c. Observes the students during the actual crossing to insure that all safety procedures are followed. It may be necessary for the observer-instructor to personally supervise the crossing even though this detracts from the realism of the problem.

4. Method of Establishing a Rope Bridge

a. The student patrol leader designates the two best swimmers as lifeguards during the crossing. If possible, these two men are qualified lifeguards. He also designates one other man to assist the lifeguards. This assistant must also be a good swimmer. These men are numbered 1, 2, and 3. Numbers 1 and 2 are the lifeguards, and number 3 the assistant.

b. Numbers 1 and 2 strip to their shorts. Their clothing and equipment are carried across the river by other students.

c. The patrol leader should use the rope provided by supply or, in its absence, he should collect and connect enough individual survival ropes to reach across the river. Number 1 lifeguard then positions himself 10 yards downstream from the crossing site. Number 2 ties one end of the small rope around his waist, using a quick release, nonslip knot. He then walks slowly into the water, checking for snags and floating debris, and tows the rope to the far bank. Number 3 feeds out the small rope, keeping it untangled and free of snags.

d. When the No. 2 man reaches the far bank, No. 3 then ties his end of the small rope to the larger rope. Number 2 pulls the larger rope across and secures it to a tree five to ten feet back from the bank. Number 3 man secures his end of the large rope in the same manner. Number 2 man then takes a position downstream from the crossing site opposite the No. 1 lifeguard.

5. Conduct of Crossing Using Rope Bridge

a. The students cross with equipment fixed to their suspenders and cartridge belts. Cartridge belts are not buckled. Rifles are slung over either shoulder with the sling tied with quick release knots to the suspenders by leather thongs or string. In case of an emergency, the student should be able to drop his suspenders, rifle, cartridge belt, and equipment in one quick move. The students cross the river by sliding both hands on the rope, hand to hand. Both hands are kept on the rope at all times. The students face upstream to watch for floating debris. The number of students on the rope at any one time should not exceed the number of life guards present.

b. Heavy equipment such as machine guns or radios should be secured to the rope by means of survival ropes in such a manner that it can be slid along the rope when pushed or pulled.

c. The No. 1 lifeguard is the last member of the patrol to cross. Before crossing, he unties the rope and secures it around his waist in a quick release, nonslip knot. Number 2 lifeguard maintains his position until No. 1 is safely across the river.

d. After everyone crosses, the lifeguards dress and the other men secure their ropes and equipment.

6. Variation in Conduct of Crossing

If the patrol leader decides to have his men carry their clothing across or float equipment on brush rafts, such variations should be allowed, if possible. In all cases, however, the rope should be placed and lifeguards posted as outlined in paragraph 4. Observer-instructors should use their judgment as to whether the variation is safe. He modifies the procedure if he deems it necessary.

7. Small Boat Safety Precautions

a. When using small boats, all students attach their equipment to their suspenders and cartridge belts only when life preservers are not available. Rifles are slung (sling loose) across the individual's back with the fast release catch of the sling located in the front so that it can be quickly operated.

b. When using life preservers, students wear them over suspenders and cartridge belts. Student equipment is secured to the suspenders and cartridge belts. Rifles are carried as outlined in subparagraph *a*, above. Rifles should be slung over the life preserver.

c. Observer-instructors insure that the students use safe procedures in embarking and debarking from the boats. They determine prior to embarking which students are weak or non-swimmers. These studens are placed in boats in which there are strong swimmers.

8. Lifesaving Procedure

In lifesaving, physical contact is a last resort. Normally, the best method of lifesaving, when boats are not available, is accomplished by extending some object such as a stick, rifle, belt, rope, or shirt, to the victim.

APPENDIX III
REVIEW TRAINING

Program of Instruction

Subject	Hours	
	Day	Night
Orientation	1	0
Intelligence	2	0
Patrol order and reports	4	0
Hand to hand combat	7	0
Physical conditioning	13	0
Tactical training	9	4
Demolitions, mines and booby traps	16	0
Map reading, aerial photograph reading, and field sketching	17	8
Ambush and roadblock techniques	2	0
Patrol tips	2	0
Aerial resupply	4	0
Communication	2	0
Ropes and knots	2	0
Total	81	12
Special Subjects	7	0
Total	88	12

1. Orientation
(1 hr)

PURPOSE: To provide a clear understanding of the purpose and importance of ranger type training. To orient the individual on the background of the American Ranger and the present concept of training.

Subject	Hours and type of instruction	Scope of instruction	Reference
Orientation	1-L, TF	LECTURE on the importance of ranger training and give the background of rangers in the American Army. Outline the training that will be undertaken by the individual with his unit. Show Film Bulletin 283, "Ranger Training."	Appendix VI, this manual.

2. Intelligence
(2 hrs)

PURPOSE: To familiarize the individual soldier with the importance of intelligence. To show his responsibilities for collecting, recording and forwarding information so that it may be processed into intelligence. To familiarize him with the procedures for processing captured enemy personnel, documents, and equipment.

Subject	Hours and type of instruction	Scope of instruction	Reference
Intelligence procedures for the individual solider lesson No. 1.	1–C	CONFERENCE on the importance of combat intelligence with particular emphasis on the individual soldier's part in collecting and transmitting this information.	FM 21–75, ASubjScd 21–7.
Handling POW's, Enemy Documents, and Captured Equipment Lesson No. 2.	1–C, TF	CONFERENCE on method employed by the individual soldier in processing POW's, enemy documents, and equipment, to include emphasis on searching, methods of segregating, and guarding prisoners. TF 30–1493, "POW's for Intelligence."	FM 21–75, FM 30–15, ASubjScd 21–7.

3. Patrol Order and Reports
(4 hrs)

PURPOSE: To provide specific training in the preparation of patrol orders and patrol reports. To familiarize the soldier with the formats for the patrol order and patrol report. To emphasize the importance of a complete order and the value of a good patrol report.

Subject	Hours and type of instruction	Scope of instruction	Reference
Patrol Order Lesson No. 1	2-C, PE	CONFERENCE emphasizing the importance of patrol orders, acquainting the soldier with the form for a patrol order and the techniques used in its preparation. PRACTICAL EXERCISE in the preparation of a patrol order.	FM 7-10, ASubjScd 7-9.
Patrol Reports Lesson No. 2	2-C, PE	CONFERENCE on the techniques and methods of reporting results and information obtained by patrol action. PRACTICAL EXERCISE in patrol reporting, to include debriefing.	FM 21-75, ASubjScd 7-9.

4. Hand to Hand Combat
(7 hrs)

PURPOSE: To develop a fundamental ability in the art of hand to hand combat so that these basic techniques may later be more fully developed by the individual through further practice and application. A means of inculcating a spirit of aggressiveness into the training program and furthering the physical fitness of the individual.

Subject	Hours and type of instruction	Scope of instruction	Reference
Hand to Hand Combat Lesson No. 1.	1–C, D, PE	CONFERENCE to include introduction to the art of unarmed combat, use of balance, internal oblique muscles, and movement; vulnerable parts of man's body, fall position, hip throw, and overshoulder throw. DEMONSTRATION to include proper fall positions for hip and overshoulder throws and proper execution of the hip throw and overshoulder throw.	FM 21–150.

Hand to Hand Combat Lesson No. 2.	1–C, D, PE	PRACTICAL EXERCISE to include conditioning exercises, proper fall position for the hip throw and overshoulder throw, and the execution of the hip throw and overshoulder throw. CONFERENCE to include introduction to the overhead throw, reverse hip throw, and rear strangle takedown. DEMONSTRATION to include overhead throw, reverse hip throw, and rear strangle takedown. PRACTICAL EXERCISE to include review of previous instruction conditioning exercises, executing overhead throw, reverse hip throw, and rear strangle takedown.	FM 21–150.
Hand to Hand Combat Lesson No. 3.	1–C, D, PE	CONFERENCE to include introduction to methods of silencing sentries and variations of basic hip and overshoulder throws.	FM 21–150.

Subject	Hours and type of instruction	Scope of instruction	Reference
Hand to Hand Combat Lesson No. 3—Continued.		DEMONSTRATION to include methods of silencing sentries and variations of basic hip and overshoulder throws. PRACTICAL EXERCISE to include review, execution of methods used to silence sentries, and execution of variations of basic hip and overshoulder throws.	
Hand to Hand Combat Lesson No. 4.	1–C, D, PE	CONFERENCE to include introduction to counters and takedowns. DEMONSTRATION to include counters and takedowns. PRACTICAL EXERCISE to include review, applying counters and takedowns, and conditioning exercises.	FM 21–150.
Hand to Hand Combat Lesson No. 5.	1–C, D, PE	CONFERENCE to include introduction to knife disarming. DEMONSTRATION to include knife disarming.	FM 21–150.

Hand to Hand Combat Lesson No. 6.	1-C, D, PE	PRACTICAL EXERCISE to include conditioning exercises, review, and student participation in knife disarming. CONFERENCE to include introduction to methods of taking prisoners and use of the knife as an offensive weapon. DEMONSTRATION to include method of taking prisoners and use of the knife as an offensive weapon. PRACTICAL EXERCISE to include review, conditioning exercise, participation in methods of taking prisoners, and use of knife as an offensive weapon.	FM 21-150. FM 21-150. FM 21-150.
Hand to Hand Combat Lesson No. 7.	1-PE	PRACTICAL EXERCISE to include review, conditioning exercises, and participation in selected exercises presented in previous periods of instruction.	FM 21-150.

5. Physical Conditioning
(13 hrs)

PURPOSE: To develop the physical fitness and mental alertness of the individual soldier.

Subject	Hours and type of instruction	Scope of instruction	Reference
Physical Conditioning Lesson No. 1.	1–C, D, PE	Integrated CONFERENCE DEMONSTRATION, and PRACTICAL EXERCISE, to introduce exercises 6, 8, 9 of Drill No. 1, two minute sit-ups, pull-ups, and 100-yard dash. CONFERENCE to include introduction to exercises.	FM 21–20.
		DEMONSTRATION of proper method of executing exercises. PRACTICAL EXERCISE to include six repetitions of exercises 6, 8, 9 of Drill No. 1, execution of two minute sit-ups, pull-ups, and 100-yard dash.	FM 21–20.

Physical Conditioning Lesson No. 2.	1–C, D, PE	Integrated CONFERENCE DEMONSTRATION, and PRACTICAL EXERCISE to introduce exercises 6, 8, 9 of Drill No. 1, two minute sit-ups, pull-ups, and ½-mile run. DEMONSTRATION of proper method of executing exercises. PRACTICAL EXERCISE to include six repetitions of exercises 6, 8, 9 of Drill No. 1, execution of two minute sit-ups, pull-ups, and ½-mile run.	FM 21–20.
Physical Conditioning Lesson No. 3.	1–PE	PRACTICAL EXERCISE to include seven repetitions of exercises 6, 8, 9 of Drill No. 1, sit-ups, pull-ups, and a ½-mile run w/M–1 rifle.	FM 21–20.
Physical Conditioning Lesson No. 4.	1–PE	PRACTICAL EXERCISE to include eight repetitions of exercises 6, 8, 9 of Drill No. 1, sit-ups, pull-ups, and a ¾-mile run.	FM 21–20.
Physical Conditioning Lesson No. 5.	1–PE	PRACTICAL EXERCISE to include eight repetitions of exercises 6, 8, 9 of Drill No. 1, sit-ups, pull-ups, and a ¾-mile run.	FM 21–20.

Subject	Hours and type of instruction	Scope of instruction	Reference
Physical Conditioning Lesson No. 6.	1-PE	PRACTICAL EXERCISE to include ten repetitions of exercises 6, 8, 9 of Drill No. 1, sit-ups, pull-ups, and a ¾-mile run w/M-1 rifle.	FM 21-20.
Physical Conditioning Lesson No. 7.	1-PE	PRACTICAL EXERCISE to include eleven repetitions of exercises 6, 8, 9 of Drill No. 1, sit-ups, pull-ups, and a 1-mile run.	FM 21-20.
Physical Conditioning Lesson No. 8.	1-PE	PRACTICAL EXERCISE to include twelve repetitions of exercises 6, 8, 9 of Drill No. 1, sit-ups, pull-ups, and a 1-mile run.	FM 21-20.
Physical Conditioning Lesson No. 9.	1-PE	PRACTICAL EXERCISE to include warmup exercises and a tug of war contest between squads and platoons of company.	FM 21-20.
Physical Fitness Test Lesson No. 10.	4-C, D, PE	Integrated CONFERENCE DEMONSTRATION, and PRACTICAL EXERCISE.	FM 21-20.

	CONFERENCE to include discussion of all phases of physical fitness test. DEMONSTRATION of proper method of executing physical fitness test. PRACTICAL EXERCISE to include actual running of the physical fitness test.

6. Tactical Training—Individual, Squad, and Patrol
(13 hrs)

PURPOSE: To teach the techniques involved in the application of basic tactical principles in individual, squad, and patrol operations. To teach by basic practical exercises the planning, preparation, and application of these principles and techniques to insure that the individual is prepared to function effectively under combat conditions.

Subject	Hours and type of instruction	Scope of instruction	Reference
Combat Formation, Rifle Squad and Platoon Lesson No. 1.	2–C, PE, TF	CONFERENCE and PRACTICAL EXERCISE on combat formation of the rifle squad and platoon. CONFERENCE to include arm and hand signals and other signals commonly used. Organization of rifle squad and platoon, the tactical advantages and disadvantages of combat formations used by the squad and platoon, and the application of prescribed formations when subjected to enemy fire. PRACTICAL EXERCISES to include breakdown into squads and platoons and execution of the formation upon command by the leader. TF 7–1919, parts 1 and 2, "Combat Formations, Rifle Squad and Platoon."	FM 7–10, ASubjScds 7–2, 7–20, 7–30, 21–1, 21–20, and 21–25.

Individual Day Training Lesson No. 2.	4-C, D, PE	CONFERENCE, DEMONSTRATION, and PRACTICAL EXERCISE on individual day training. CONFERENCE and DEMONSTRATION to include proper individual training and camouflage and the use of cover and concealment. Methods of movement and selecting routes. Maintenance of direction, observation techniques, and eye and ear training. PRACTICAL EXERCISE to include application of the principles discussed and demonstrated during the conference.	FM21-75, ASubjScds 7-2, 7-20, 7-30, 21-1, 21-20, and 21-25.
Individual Night Training Lesson No. 3.	4-C, D, PE	CONFERENCE, DEMONSTRATION, and PRACTICAL EXERCISE on individual night training. CONFERENCE and DEMONSTRATION to include dress of the individual, action and control as a member of a patrol, wire cutting, action under flares, means of identification of individuals, maintenance of direc-	FM21-75, ASubjScds 7-2, 7-20, 7-30, 21-1, 21-20, and 21-25.

Subject	Hours and type of instruction	Scope of instruction	Reference
Individual Night Training Lesson No. 3—Continued.		tion by use of compass, stars, terrain, and eye and ear training. PRACTICAL EXERCISE will emphasize the principles discussed and demonstrated during the conference.	
Troop Leading Procedures Lesson No. 4.	2–C	CONFERENCE to include logical sequence of thought and action in preparing an order for a combat operation, methods of coordinating and conducting reconnaissance, formulating a plan, and issuing the order.	FM 7–10, ASubjScds 7–2, 7–20, 7–30, 21–1, 21–20, and 21–25.
Evasion and Escape Lesson No. 5.	1–C	CONFERENCE on evasion and escape, methods to be employed by the individual soldier while in enemy territory, including planning, reconnaissance, evasion tactics, methods of escape and special techniques concerning survival while escaping and evading.	DA Pam 21–46.

7. Demolitions, Mines, and Booby Traps
(16 hrs)

PURPOSE: To develop a working proficiency with combat military explosives and accessory demolition items with emphasis on nonelectric firing; a familiarization with US AT and AP mines; and a familiarization with all standard firing devices for booby traps. To accomplish the maximum in practical work along with minimum conference during all phases of demolitions and booby trap training. To require individuals to perform practical work in the calculation, preparation, priming, detonation, and neutralization of assault type charges and booby traps.

Subject	Hours and type of instruction	Scope of instruction	Reference
Introduction to Military Demolitions Lesson No. 1.	4–C; D, PE	CONFERENCE, DEMONSTRATION and PRACTICAL EXERCISE on combat military explosives and accessories. CONFERENCE and DEMONSTRATION to include description and characteristics of the four best military explosives, time fuse, detonating	FM 5–25, TM 9–1900, ASubjScd 5–6.

Subject	Hours and type of instruction	Scope of instruction	Reference
Introduction to Military Demolitions Lesson No. 1—Continued.		cord, fuse lighters, nonelectric blasting caps, priming adapters, and detonating cord clips. PRACTICAL EXERCISE to include the assembling of explosives with various primers and demolition ties, the technique of crimping, and the makeup and firing of a nonelectrical charge.	
Calculation and Placement of Demolitions Lesson No. 2.	4–C, D, PE	CONFERENCE to include calculation and placement of steel, concrete, and cratering barrier charges, neutralization of explosive charges, and electrical firing systems. DEMONSTRATION to include placement of various charges. Actual firing of nonelectrical charges. PRACTICAL EXERCISE to include calculation, placement, firing of charges nonelectrically, and makeup	FM 5–25, TM 9–1900, ASubjScd 5–6.

Demolition Projects Lesson No. 3.	4-C, D, PE	and detonation of electrical charges in series. CONFERENCE to include use of assault charges, bangalore torpedoes, shaped charges, pole and satchel charges, calculation formula, placement of external nonelectrical timber cutting charges. DEMONSTRATION to include assembly of various charges, firing of charges and bangalore torpedoes. PRACTICAL EXERCISE to include makeup and firing of pole charges; priming and firing of shaped charges; priming and firing of bangalore torpedoes and tree cutting external charges.	FM 5-25, TM 9-1900, ASubjScd 5-6.
Mines and Booby Traps Lesson No. 4.	4-C, D, PE	CONFERENCE to include types, terminology, and uses of US antipersonnel mines and booby traps. DEMONSTRATION to include nomenclature and functioning, and activation and deactivation procedures for all US standard fuses.	TM 9-1900, FM 5-32, ASubjScd 21-23.

Subject	Hours and type of instruction	Scope of instruction	Reference
Mines and Booby Traps Lesson No. 4—Continued.		PRACTICAL EXERCISE, to include arming and disarming fuses, mines, and booby traps.	

8. **Map Reading, Aerial Photograph Reading, and Field Sketching** (25 hrs)

PURPOSE: To provide a review of the basic principles and techniques in map reading, aerial photograph reading and field sketching that the soldier may normally be expected to apply as a small unit leader in combat. To develop in the individual a working knowledge of the basic principles of field navigation and, by application of these principles, to develop confidence in his abilities.

Subject	Hours and type of instruction	Scope of instruction	Reference
Map Reading Lesson No. 1	4-C, PE	CONFERENCE covering map reading to include marginal information, topographical and military symbols, military grid coordinates, map orientations, scale, direction and azimuth, declination diagram, G-M angle, polar coordinates, relief and topography, profiles, intersection, and modified resection. PRACTICAL EXERCISE on each topic covered above.	FM21-26, FM21-30, ASubjScd 21-9, and 30-15.
Aerial Photograph Reading Lesson No. 2.	2-C, PE	CONFERENCE covering types of aerial photos including basic photo interpretation, point designation grids, scale, and method of determining magnetic North. PRACTICAL EXERCISE on each topic covered above.	FM21-26, ASubjScd 21-9, and 30-15.

Subject	Hours and type of instruction	Scope of instruction	References
Map and Aerial Photograph Reading Lesson No. 3.	4–C, PE	CONFERENCE on methods of determining north by use of a watch, the sun, and the stars; explanation of the use of compass and methods used to set compass for day or night reading. PRACTICAL EXERCISE to include field applications of major principles of map and aerial photo reading, including orientation of map and photos, terrain study, use of the compass, location of ground position on map and photo by inspection, location of ground position by resection, location of ground feature by intersection and a short cross country compass march.	FM 21–25, FM 21–26, ASubjScd 21–9, and 30–15.
Map and Aerial Photograph Reading Lesson No. 4.	12–PE	PRACTICAL EXERCISE in the field on map and aerial photo reading during daylight and darkness. Requires individuals to reach ten differ-	FM21–25, FM21–26, ASubjScd 21–9, and 30–15.

Field Sketching Lesson No. 5	3-C, D, PE	ent stations using topographical maps, photo maps, compass, and proper pacing. Participation to require map orientation, use of military symbols, and use of PD, and military grid coordinates. CONFERENCE to include the basic fundamentals necessary in expedient field sketching of terrain, military objectives, installations, and sites peculiar to the information normally desired for a reconnaissance patrol. Emphasis to be placed on area sketching and panoramic sketching. DEMONSTRATION to include fundamentals of field sketching. PRACTICAL EXERCISE to include student application of above.	FM 21-35, ASubj-Scd 21-9, and 30-15.

9. Ambush and Roadblock Techniques
(2 hrs)

PURPOSE: To familiarize the student with techniques employed in ambushing enemy personnel and motorized convoys. To show various methods and principles involved in establishing roadblocks. To show method of establishing the roadblock as an adjunct to the ambush.

Subject	Hours and type of instruction	Scope of instruction	Reference
Ambush and Roadblock Techniques.	2–C, D	Integrated CONFERENCE and DEMONSTRATION showing techniques and methods involved in conducting ambushes and establishing roadblocks.	Chapter 3, this manual.

10. Patrol Tips
(2 hrs)

PURPOSE: To familiarize the student with the various tips and techniques used in conjunction with patrolling operations.

Subject	Hours and type of instruction	Scope of instruction	Reference
Patrol Tips	2–C, 1D	Integrated CONFERENCE and DEMONSTRATION to familiarize the student with the various patrol tips and techniques.	App. V, this manual.

11. Aerial Resupply
(4 hrs)

PURPOSE: To familiarize the student with the methods of aerial resupply, the coordination and planning for aerial resupply, and the technique of organizing a DZ behind enemy lines.

Subject	Hours and type of instruction	Scope of instruction	Reference
Familiarization with Army Aircraft and Aerial Resupply Lesson No. 1.	1-C, D	CONFERENCE and DEMONSTRATION to familiarize the students with the command channels for requesting aerial resupply, equipment used for aerial resupply, selection and marking of DZ's or LZ's, containers and preparation of equipment for airdrop.	Chapter 2, this manual; FM 57-35.
Aerial Resupply Lesson No. 2	3-C, PE, D	CONFERENCE to include instruction on organization of patrol to receive airdrop, preparing and issuing annex to patrol order covering aerial resupply, action upon receipt of airdrop, to include recovery and redistribution. PRACTICAL EXERCISE to include writing and issuing aerial resupply annex to patrol order. DEMONSTRATION to include the tactical organization of a patrol on	ASubjScd 7-53, 7-54.

	DZ and the action of patrol during helicopter resupply and evacuation of letter; resupply by parachute; and free fall by light and cargo-type aviation.

12. Communication
(2 hrs)

PURPOSE: To acquaint and familiarize the student in radio telephone procedures of radio AN/PRC-10 and AN/PRC-6, to include practical exercise in setting up for operation and calibration.

Subject	Hours and type of instruction	Scope of instruction	Reference
Communication Lesson No. 1.	1-C, D	Integrated CONFERENCE, DEMONSTRATION and familiarization instruction of radiotelephone procedure, to include establishing communication, organization of radio net, transmission of messages, and transmission security.	ASubjScd 21-13

Subject	Hours and type of instruction	Scope of instruction	Reference
Communication Lesson No. 2.	1–C, D, PE	Integrated CONFERENCE, DEMONSTRATION and PRACTICAL EXERCISE, to include familiarization instruction in the operation of radio set AN/PRC-10 and AN/PRC-6. DEMONSTRATION to include installation of battery operation in internal and external positions for radio set AN/PRC-6. DEMONSTRATION and PRACTICAL EXERCISE to include installation of battery and setting up for operation to include calibration.	ASubj Scd 21-13.

13. Ropes and Knots
(2 hrs)

PURPOSE: To familiarize students with the types of rope and the practical use of each type as related to traversing difficult terrain which may be encountered in ranger operations. To give students a practical knowledge in the tying and proper use of various knots.

Subject	Hours and type of instruction	Scope of instruction	Reference
Ropes and Knots	2-C, D, PE	CONFERENCE to include use of ropes and knots in relation to ranger operations. DEMONSTRATION to include types of ropes, tying, and use of basic knots. PRACTICAL EXERCISE to include tying of all basic knots.	FM 70-10.

14. Special Subjects

PURPOSE: To provide necessary instruction in subjects applicable to the terrain peculiar to a unit's location and climatic conditions. To equip the individual with a working knowledge of the principles and techniques he will be required to apply in the field training of his unit.

Subject	Hours and type of instruction	Scope of instruction	Reference
Survival	As desired	Instruction in survival techniques to include methods of recognizing dangers inherent in local area. Identification of dangerous ground, poisonous plants, reptiles, and animals. Instruction to include identification of edible plant and animal life. Instruction on means of catching and preparing wild life and how to construct shelters.	Appropriate to location.
Fieldcraft Expedients	As desired	Instruction on general field expedients that may be used to maintain direction, overcome obstacles, or help alleviate discomfort in difficult terrain. See FM 72–20, Jungle Operations; FM 70–10, Mountain Operations; FM's in the 31-series, Special Operations, as appropriate.	Appropriate to location.

Mountain Techniques and Expedients.	Instruction on techniques and expedients peculiar to operations in mountainous terrain. Instruction on how to conserve energy by pacing oneself during movement in the mountains and how to overcome obstacles.	FM 70-10.
Desert Techniques and Expedients.	Instruction on techniques and expedients peculiar to operations in desert terrain. Instruction on how to conserve energy by pacing oneself during movement in the desert and how to overcome obstacles.	FM 31-25.
Jungle Techniques and Expedients.	Instruction on techniques and expedients peculiar to operations in jungle terrain. Instruction on how to conserve energy during movement in the jungle and how to overcome obstacles.	FM 72-70.
Cliff Assault Techniques	Instruction on techniques of assaulting cliffs, to include beach landings on hostile shores, scaling techniques, beach, and cliffhead security.	Chapter 4 this manual.

SAMPLE SCHEDULE
1st Week

Day	1st period	2d period	3d period	4th period	5th period	6th period	7th period	8th period	Night
Monday	Orientation	Commanders Time—Special Subjects			Troop Leading Procedures		Hand to Hand Combat	PT	
Tuesday	Combat Formations			Demolitions			Hand to Hand Combat	PT	
Wednesday	Patrol Order			Map Reading			Hand to Hand Combat	PT	
Thursday	Aerial Photography			Demolitions			Hand to Hand Combat	PT	
Friday	Patrol Report			Map Reading			Hand to Hand Combat	PT	
Saturday	Aerial Resupply								

SAMPLE SCHEDULE
2d Week

Day	1st period	2d period	3d period	4th period	5th period	6th period	7th period	8th period	Night
Monday	PT	Sketching			Map Reading				To 0100
Tuesday	Open			Demolitions			Hand to Hand Combat	PT	
Wednesday	Individual Day Training				Intelligence		Hand to Hand Combat	PT	Individual Night Training 1900–2300
Thursday	Open	Escape and Evasion	Mines and Booby Traps				PT	Open	
Friday	Physical Fitness Test				Patrol Tips		Communication		
Saturday	Ambush and Road Block Technique		Ropes and Knots						

APPENDIX IV
FIELD TRAINING EXERCISES

Section I. INTRODUCTION

1. Purpose and Scope

a. This appendix contains field exercises presented during the last three weeks of ranger training. Ten exercises are included. These problems are effective in developing the capabilities and leadership of platoon and squad leaders and their subordinates, but all personnel of the participating unit receive invaluable benefits.

b. This three weeks' training period is the closest approximation to combat conditions that can be achieved outside of actual combat. During this cycle there is no provision for free time; therefore, training continues through weekends and holidays. Training is never postponed because of inclement weather. The training is actually the same as a unit experiences while undergoing field maneuvers or while engaged in combat. The men experience the fatigue, frustration, and mental strain that is ever present in combat.

c. Army Subject Schedules should be used to assist in the organization of the material contained in each field exercise. These Subject Schedules are referenced in appendix I.

PROGRAM OF INSTRUCTION

Exercise	Hours
Section II. Day reconnaissance of enemy battle position	10
III. Night raid on enemy position to obtain prisoners	15
IV. Night raid to destroy enemy outpost	12
V. Night raid against enemy rear area installation	19
VI. Night infiltration and reconnaissance of an area deep in enemy territory	48
VII. Raid against enemy guerilla camp	24
VIII. Raid to seize and hold key enemy installation	36
IX. Night infiltration and ambush	26
X. Waterborne raid against a critical installation	45
XI. Raid against installation deep in enemy territory	72

Section II. DAY RECONNAISSANCE OF ENEMY BATTLE POSITION

2. Purpose

To develop the ability of leaders to prepare a thorough patrol plan and execute it under realistic combat conditions; to familiarize the members of the patrol with the principles and techniques of patrolling; and to develop the ability to utilize stealth, cover, and concealment to move near the enemy's positions.

3. Scope*

a. A day reconnaissance patrol exercise to include planning, preparation, execution, and critique phases. The problem requires obtaining information of the

---

* Subparagraphs *b, c, d,* and *e* of paragraph 3 are also applicable to all of the field exercises in this text. They have been deleted from the following exercises to avoid repetition.

enemy's battle position. The mission is to be accomplished over areas under observation by the enemy, thus necessitating maximum use of stealth, cover, and concealment.

b. Planning phase to stress patrol leader's actions upon receipt of order, his issuance of a warning order, and selection of equipment to accomplish the patrol mission.

c. Preparation phase to evaluate leader's organization of necessary preparatory measures, reconnaissance, issuance of patrol order, supervision of patrol preparation, inspection of patrol, and rehearsal. To evaluate the completeness and soundness of his plan. To evaluate subordinate leaders and members of the patrol in their performance of assigned preparatory tasks. To observe their attention to details, understanding of the situation, and teamwork.

d. Execution phase to evaluate leader's ability to execute his plan and to observe his initiative, decisiveness, and aggressiveness. Evaluate the attitude, cooperation, and actions of subordinate leaders and patrol members during the execution of the problem.

e. Critique phase to review the principles and techniques of patrolling involved in the exercise. To bring out the correct techniques used on the patrol and, by discussion, to correct errors committed.

f. See FM 21-75.

4. Scenario

a. This problem should be used as an introductory exercise to familiarize and develop the individual and the unit for the longer and more difficult problems that follow. Although it is short, both in distance to be covered and the length of time required, it contains the requirements necessary in the planning or preparation phase and sufficient subsequent requirements to familiarize personnel with the techniques and principles involved in the execution of any ranger type field exercise. The size of the patrol on this problem should not be more than 9 nor less than 4 men.

b. The patrol is oriented in a location designated as a reserve battle group area. This orientation should be conducted by an individual designated as the battle group S2. It includes the purpose of the problem, necessary safety and administrative details, and a presentation of the general tactical situation. Following orientation of the patrol, the patrol leader receives an operation order to include his patrol's specific mission. Maps, overlays, aerial photographs, and terrain models may be used in conjunction with the issuance of the operation order; however, the types and quantities of the visual aids employed are consistent with those normally found under similar circumstances in a combat situation.

c. The patrol leader is now responsible for the execution of the proper steps in troop leading procedure required for the successful accomplishment of his mission. Generally, his actions follow the sequence contained in the word picture of the problem presented below.

> (1) The patrol leader makes an initial estimate of the situation and decides on an efficient way to utilize the time available to him. He makes a thorough map study and comes up with a preliminary plan. He issues a warning order so that preparations can be started by all members. He makes any necessary coordination with personnel available in the

area. His coordination completed at battle group level, he departs for the forward company through which his patrol is to pass. He coordinates with the company commander and obtains any information that may be available. He then is guided to a forward outpost where he makes a visual reconnaissance of the terrain between his lines and the objective. He discusses pertinent details with the individuals in the outpost. The patrol leader now returns to his patrol. He checks the progress of the patrol's preparation and makes changes in his preliminary plan, if necessary. He then completes his plan and prepares his patrol order. At the time and place designated in his warning order, he meets with his patrol and issues the patrol order. After answering questions and satisfying himself that his men understand the mission and their respective duties, he dismisses them to complete their preparations. He supervises the patrol and conducts any rehearsals he feels are necessary, such as formations that may be used and review of signals to be used. When possible, rehearsals are conducted on terrain similar to that over which the patrol is to operate. Prior to departure time, the leader conducts a final inspection. At the designated time the patrol is moved forward to the forward company command post area. There the patrol leader makes a final check with the company's commanding officer. Then he and his patrol are guided forward to the outpost. The patrol leader checks with the personnel in the outpost for any last minute information they may have. He moves his patrol out and guides them on to his selected route. He utilizes pace men, point men, and compass men as desired. He adjusts or changes his formation based on the terrain, cover, concealment, visibility, and proximity to known enemy positions.

(2) When the patrol nears the enemy lines, the patrol leader selects a suitable area for a security group. He leaves the previously designated security group in position and moves out on his reconnaissance with the individual or individuals he has designated to accompany him. The special situation imposes a time restriction on the patrol leader. He may move freely in his assigned objective area up to a certain time. At this designated time, however, he must be clear of the area because friendly artillery is shifting back in on the objective. His reconnaissance completed, the patrol leader and his reconnaissance group rejoin the security group. At this point the information obtained is passed on to all members of the patrol. The patrol now moves out on its return route. Approaching the designated point of return in the friendly lines, the patrol leader slows down the patrol and moves cautiously until contact is established and recognition accomplished. The patrol passes through the friendly unit and returns to its area. The patrol leader makes an immediate check of all personnel to see if

anyone picked up any information that should be included in his report.

(3) In order to provide training for all personnel, the patrol leader may be debriefed in the presence of the entire patrol. The patrol leader makes his report and the observer-instructor, who was with the patrol throughout the problem, conducts a critique. Following the critique the observer-instructor discusses leadership and command weaknesses with each individual as he deems necessary.

(4) The observer-instructor accompanying the patrol constantly observes the actions of all members of the patrol. He assesses casualties, as explained in chapter 5; he does not interfere with the problem in any way except in an emergency.

5. General and Special Situation

A general and special situation constituting an operation order is presented by the principal instructor who introduces himself as the S2. The operations order may be given in its correct sequence, or the situation may be presented in a series of notes not necessarily in correct operations order sequence. The latter method is effective in training an individual leader in how to rearrange his notes into the proper sequence for presentation as his patrol order. The following is a suggested method of presenting the situation for this problem:

a. General Situation. Suspected enemy positions are shown on acetate overlay. Exact position unknown. Morale unknown. Unit believed to be _____ Aggressor division. The Aggressor has tanks, mortar, and artillery support. Air has observed considerable activity on Aggressor's MLR's.

  b. *Special Situation.*

   (1) Your company has been directed to send out reconnaissance patrols to determine the strength, activity, and disposition of the enemy in the battle group sector.

   (2) Your mission is to reconnoiter the high ground in your assigned sector; the specific coordinates will be given you later. You will note the terrain both on your way out and return route.

   (3) You will be taken to the forward company through which you will depart in time to make a visual reconnaissance of the company area and the terrain over which the patrol will move.

   (4) You will cross the LD at ____ hours.

   (5) The latest information will be given you at the company command post.

   (6) The ____ field artillery battallion will be firing on your objective until ____ hours today. It will resume firing at ____ hours.

   (7) Return through your friendly outpost by ____ hours.

   (8) The coordinates of your specific objectives are ____.

   (9) Casualties will be evacuated through the friendly outpost from which you depart. The patrol leader will determine the disposition to be made of all casualties.

   (10) One "C" meal will be carried by each man.

(11) Battle group will furnish transportation. Draw all supplies and ammunition from your company.

(12) I will receive your patrol report here when you return.

(13) No radios will be carried.

(14) The challenge until ____ hours is _____. The password is _____. The time is ____.

(15) The weather is _____. Darkness is at (time).

(16) Are there any questions?

(17) The time is now ____.

  c. Critique Checklist.

    (1) Was a good plan made for the reconnaissance, and was the reconnaissance conducted properly?

    (2) Was information obtained on reconnaissance passed on to all members of the patrol?

SUGGESTED TIME SCHEDULE

0800–0820	Briefing.
0820–0830	Prepare and issue warning order.
0830–0845	Coordination in reserve area.
0845–1015	Travel to forward unit—coordinate—make visual reconnaissance—return to reserve area.
1015–1115	Prepare patrol order.
1115–1300	Issue order—supervise final preparations—chow—rehearsal—inspection.
1300–1330	Move to line of departure.
1330–1630	Depart OP (LD)—move to objective—begin return when reconnaissance is

	complete but not later than 1630 due to shift of friendly artillery onto objective.
1630–1745	Return to friendly lines.
1745–1800	Return to reserve area.
1800–2000	Debriefing—chow—critique.

SKETCH OF PROBLEM LAYOUT

Enemy Battle Positions
———————————
↕ 1,800–2,000 yds

Friendly Company Outpost (L.D.)
———————————
↕ 50–200 yds

Friendly Company Command Post
———————————
↕ As desired

Reserve Area

No special features required

Note. It is recommended that the distance for movement for a patrol on this problem be approximately 4,000 yards. However, the distance can be scaled to suit the available terrain. A sufficient number of situations should be included to keep the patrol occupied for a period of five hours after departing the outpost.

Section III. NIGHT RAID ON ENEMY LINES TO OBTAIN PRISONERS

6. Purpose

To develop the ability of leaders to prepare a thorough patrol plan and to execute the plan under realistic combat conditions. To familiarize the members of the patrol with the principles and techniques of night patrolling. To develop the ability to seal off a portion of the enemy's battle position by use of supporting fire and to move into the sealed off area, attack his positions, capture prisoners, and withdraw.

7. Scope

a. Night combat patrol. A night combat patrolling exercise to include planning, preparation, execution, and critique phases. The problem requires the use of support fires to seal off a portion of the enemy's battle position, movement into the sealed off area, conduct of a raid on his positions, capture of prisoners, and return to friendly lines. See paragraph 3.

b. See FM 21-75.

8. Scenario

a. This problem is a relatively short night combat patrol and should be scheduled during the early phases of training. It tests the ability of the patrol leader to use supporting fires, to conduct a raid, and to capture prisoners. The size of the patrol should not be more than 30 nor less than 12 men.

b. The patrol is oriented, after which the patrol leader is given an operation order, to include his patrol's specific mission. The patrol leader is now responsible for the execution of the proper steps in troop leading procedure required for the successful accomplishment of his mission. Generally, the actions of the patrol leader follow the sequence contained in the word picture of the exercise presented below.

    (1) The patrol leader makes an initial estimate of the situation and decides on an efficient way to utilize the time available to him. He makes a thorough map study and comes up with a preliminary plan. He issues a warning order so that preparations can be started by all members. He makes any necessary coordination with personnel available in the area. His coordination completed at battle

group, he and his selected subordinates depart for the forward company through which his patrol is to pass. He coordinates with the company commander and obtains any information that may be available. He then is guided to a forward outpost where he makes a visual reconnaissance. He discusses pertinent details with the individuals in the outpost. The patrol leader and his subordinates now return to his patrol. He makes changes in his preliminary plan if necessary and checks the progress of the men's preparation. The patrol leader then completes his plan and prepares his patrol order. At the time and place designated in his warning order, he meets with his patrol and issues the patrol order. After answering questions and satisfying himself that his men understand the mission and their respective duties, he dismisses them to complete their preparations.

(2) The patrol leader supervises the patrol and conducts any rehearsals he feels are necessary, such as actions at the objective and at danger areas. When possible, rehearsals are conducted on terrain similar to the terrain in which the patrol is to operate. Prior to departure time the leader conducts a final inspection. At the designated time the patrol is moved to the forward company command post area. The patrol leader makes a final check with the company commanding officer; and then he and his patrol are guided forward to the outpost. The patrol leader checks with the personnel in the outpost for any last

minute information they may have. He moves his patrol out and guides them on his selected route. Just outside the friendly position he halts the patrol, establishes security, and listens for signs of a possible enemy ambush. He utilizes pace men, point men, and compass men as desired. He adjusts or changes his formation based on the terrain, cover, concealment, visibility, and proximity to known enemy positions. As the patrol approaches the Aggressor MLR the patrol leader insures that the patrol utilizes stealth to the maximum in order to avoid compromising the patrol's position to the enemy. Upon arriving at the assault position, the patrol leader has each man get into his proper position and then proceeds to call in his prearranged artillery fire on the objective area. This is accomplished through the observer-instructor, who, in this exercise, represents a forward observer. The observer-instructor carries a radio for this purpose. If the fires are coordinated and prearranged properly by the patrol leader, the Aggressor detonates demolitions on the objective position to simulate the artillery fire. If coordination is improper, the principal instructor might have demolitions fired to the right or left of the objective area in order to simulate the improper prearrangement. The patrol leader, in this case, requests through the forward observer that the artillery fire be shifted onto the objective. The demolitions on the objective are then exploded. After firing the mission the patrol leader requests that the fires be shifted to the right, left, and rear of the objective in order to seal it off. Then he launches his attack.

(3) The objective in this exercise is prepared as is the enemy battle position in chapter 5. The objective should be part of the enemy's battle position. The mission is to obtain prisoners. After this is accomplished, the patrol leader has the patrol withdraw from the position and, after clearing it, calls in artillery once again to prevent enemy countermeasures. The patrol then returns to the outpost through which it departed. Having properly given the password, it enters the outpost and begins its final movement to the reserve area. Upon arrival at the battle group reserve area, the patrol turns the prisoners over to the S2. The patrol is then debriefed, fed a hot meal, and critiqued by the observer-instructor.

9. General and Special Situation

A general and special situation constituting an operation order is presented by the principal instructor who introduces himself as the S3. The following is a suggested method of presenting the situation for this problem:

a. General Situation.
  (1) Friendly forces have been attacking generally (direction) meeting sporadic resistance. We halted at (time) hours (day) to reorganize and prepare to continue the attack on order.
  (2) Enemy action has been so sporadic that battle group does not know whether the encountered

enemy forces are part of a delaying force or part of the enemy battle position.

b. *Special Situation.*
  (1) You are presently located as shown on this overlay.
  (2) You depart this area at ____ hours.
  (3) Your mission is to capture a prisoner at the objective.
  (4) You will coordinate artillery fire in order to seal off the objective area and inflict casualties prior to your attack. You will conduct a raid on the objective located at (coordinates).
  (5) Artillery support will be furnished by the (unit).
  (6) You will be the only patrol in this sector tonight.
  (7) You will have a forward observer with you.
  (8) The friendly situation is as is shown on this overlay.
  (9) You will carry no rations.
  (10) The patrol will pass through ____ Company ____ Battle Group.
  (11) Your patrol will return prior to (time) hours.
  (12) The challenge is _____. The password is _____.
  (13) Request weapons, ammunition, and special equipment as desired.
  (14) You will take one radio AN/PRC-10. Call signs are as follows: _____, this headquarters. Channel ____. Your call sign is _____.
  (15) Evacuate casualties through the same company through which you depart.
  (16) Battle group will furnish transportation to the forward area company.

(17) I will receive your patrol report here upon your return.
(18) Weather will be _____. Darkness is at ____, daylight ____.
(19) Are there any questions?
(20) The time is now ____ hours.

c. *Critique Checklist*.
(1) Was the action at the objective well planned?
(2) Were artillery or mortar concentrations in objective area properly prearranged to include—
   (a) Initially on the objective?
   (b) Sealing off the objective during raid?
   (c) Cover of withdrawal from objective?
(3) Were prisoners captured? Were they properly handled?

SUGGESTED TIME SCHEDULE

0800–0900	Briefing.
0900–0915	Issue warning order.
0915–1200	Coordination—planning.
1200–1300	Dinner.
1300–1430	Continue planning.
1430–1630	Travel to forward area—coordinate—make visual reconnaissance—return to reserve area.
1630–1730	Issue patrol order.
1730–1830	Supper.
1830–1930	Final preparation—Rehearsal—inspection.
1930–2000	Move to forward area.
2000–2030	Move to LD.

2030–   Cross LD.
0600–   Return through friendly lines.
0600–0800   Return reserve area—debriefing—critique.

SKETCH OF PROBLEM LAYOUT

Enemy Battle Position (objective) ───────────

⋀
1,200–2000 yds
⋁

Friendly Company Outpost (L.D.) ───────────

⋀
50–200 yds
⋁

Friendly Company Command Post ───────────

⋀
As desired
⋁

Reserve Area ───────────

Note. It is recommended that the distance for movement for a patrol in this exercise be approximately 3,600 yards. However, the distance can be scaled to suit the available terrain. A sufficient number of situations should be included to keep the patrol occupied for a period of ten hours after departing the outpost.

Section IV. NIGHT RAID TO DESTROY ENEMY OUTPOST

10. Purpose

To develop the ability of leaders to prepare a thorough patrol plan and to execute it under realistic combat conditions. To familiarize the members of the patrol with the principles and techniques of night patrolling. To familiarize leaders and patrol members with the difficulties encountered in the execution of a night raid on an enemy position.

11. Scope

a. Night patrol exercise. A night combat patrolling exercise to include planning, preparation, execution, and critique phases. The problem requires the organization of a raid on an enemy outpost, the killing or capturing of all enemy personnel, and the destruction by demolitions of the position. See paragraph 3.

b. See FM 21–75 and FM 7–10.

12. Scenario

a. This exercise is short but includes much in the way of instruction. During the twelve hours allotted, there will be a briefing of the patrol, a reconnaissance, coordination with forward area positions, preparation of a patrol order and its issuance, movement through friendly forward positions, actions under enemy mortar fire, raid against an enemy position, destruction of enemy installation by live charges, action against counterattack, and movement back into friendly lines. Because of the above it is recommended as one of the introductory exercises so that the troops can become familiar with the principles and techniques to follow in longer ranger type operations. The size of the patrol in the exercise should not be more than 36 nor less than 12 men.

b. It is desirable to have a stream flowing between friendly and Aggressor positions. This stream should be fordable and about ten yards across, and the edge should be lined with trees and brush. The trees next to the stream are rigged with one-quarter pound TNT charges which are hung with marlin. These charges are fired electrically by an Aggressor located on the enemy side of the stream. The objective area should be manned by approximately 15 Aggressors. Located there is an automatic weapons bunker to be destroyed by the patrol, and a series of 2-man foxholes. Demolition pits located around the objective provide friendly artillery supporting fire for the patrol if called for.

c. Following the patrol orientation, the patrol leader receives an operation order to include his patrol's specific mission. The patrol leader now becomes responsible for the execution of the proper steps in troop leading procedure required for the successful accomplishment of his mission. Generally, the actions of the patrol leader follow the sequence contained in the word picture of the problem presented below.

    (1) The patrol leader makes an initial estimate of the situation and decides on an efficient way to utilize the time available to him. He makes a thorough map study and comes up with a preliminary plan. He issues a warning order so that preparation can be started by all members. He makes any necessary coordination with the personnel available in the area. His coordination completed at battle group, he departs for the forward company through which his patrol is to pass. He coordinates with the company commander and obtains any information that may be available. He then is guided to a forward outpost where he makes a visual reconnaissance of the terrain. He discusses pertinent details with the individuals in the outpost. The patrol leader now returns to his patrol, checks the progress of its preparation, and changes his preliminary plan if necessary. He then completes his plan and prepares his patrol order. At the time and place designated in his warn-

ing order, he meets with his patrol and issues the patrol order. After answering questions and satisfying himself that his men understand the mission and their respective duties, he dismisses them to complete their preparations. He supervises the patrol and conducts any rehearsals he feels are necessary. When possible, rehearsals should be conducted on terrain similar to that over which the patrol is to operate. Prior to departure time the leader conducts a final inspection. At the designated time, the patrol is moved forward to the forward company command post area. The patrol leader makes a final check with the company's commanding officer, and then he and his patrol are guided forward to the outpost. The patrol leader checks with the personnel in the outpost for any last minute information they may have. He moves his patrol out and guides them on to his selected route. He utilizes pace men, point men, and compass men as desired. He adjusts or changes his formation based on the terrain, cover, concealment, visibility, and proximity to known enemy positions.

(2) The observer acts as an artillery forward observer in this exercise and has an AN/PRC-10 and an artillery or grenade simulator. When the patrol reaches the stream and has progressed halfway across, the observer throws the grenade, which is the signal for the demolitions man to fire the tree charges. This represents Aggressor mortar fire. After firing the charges the Aggressor moves back to the

demolitions control board at the objective area. Here he is able to give the patrol supporting artillery fire on the objective area as requested by the patrol leader.

(3) The observer-instructor may at this time declare casualties. The patrol should move rapidly through the mortar fire, reorganize if necessary, and then move toward the objective.

(4) After making a quick reconnaissance at the objective area, the patrol leader moves his men into position to start the assault. He may request friendly artillery fire while he sets up. He makes his request through the observer-instructor.

(5) As the attack progresses, the Aggressors manning the position withdraw from the objective after offering resistance by fire. As they leave, they light up the area by igniting two trip flares tied to trees.

(6) The observer should then accompany the demolitions team of the patrol to act as safety officer and supervise the making up of the charge and its placement in the bunker. When the charge is set and burning, the patrol should start moving off of the objective back to friendly lines. This should take a very short period of time. When the charge detonates, the Aggressors who have reformed start their counterattack. If it has been ten minutes since leaving the objective and it is obvious that the patrol is confused and has not set the charge, the Aggressors counterattack. The patrol may decide to leave the

objective during this counterattack, even though it has not accomplished its mission. If the patrol decides to fight on the objective, the Aggressor is repulsed until the charge detonates. The demolition man at the control board should have saved at least eight demolition pits to be fired for realism during the Aggressor counterattack.

(7) The Aggressors may push their counterattack approximately to the stream if the patrol is in disorder.

(8) The patrol recrosses the stream and returns through the same outpost from which it departed. Correct use of the challenge and password should be noted during the return.

(9) The patrol returns to the reserve area where it is debriefed. After the debriefing and the patrol report, the observer critiques the patrol members.

13. General and Special Situation

A general and special situation constituting an operation order is presented by the principal instructor who introduces himself as the S2 or S3 as desired. The following is a suggested method of presenting the situation for this problem.

a. General Situation.

(1) The location of the Aggressor main line is at present unknown in this area. They do, however, have a line of outposts as shown on this overlay. Each outpost includes at least one automatic weapons emplacement. These outposts are not mutually supporting. Ag-

gressor morale is believed to be high even though he has had several setbacks along the line.

(2) _____ battle group of the _____ infantry division is on the left and _____ battle group on the right. This is the location of your reserve area (shown on overlay).

b. *Special Situation.*
   (1) Battle group wants the automatic weapons emplacements destroyed tonight so that they will not interfere with our attack tomorrow.
   (2) Your company will furnish a combat patrol of (number) men to pull a night raid to destroy this outpost and to kill or capture all of its occupants.
   (3) You will cross the LD at ____ hours tonight, complete your mission, and return not later than ____ hours in the morning.
   (4) You will move through _____ company of the _____ battle group. _____ company command post is located at the following coordinates _____. Your LD, which is the _____ company outpost, is located at coordinates _____.
   (5) At ____ hours a guide will take you to the forward company so you can reconnoiter the objective area by visual reconnaissance.
   (6) You will have an artillery forward observer with you on the patrol, who will give you fire support on call.
   (7) Wounded will be taken care of in the manner prescribed by the patrol leaders. You may use evacuation through ____ company.

(8) Bring back a prisoner if possible.

(9) Battle group transportation will be available at ____ hours for movement to the reconnaissance area and again at ____ hours to transport your patrol to the forward company area.

(10) Request for weapons, ammunition, and equipment will be turned in at ____ hours to your company supply.

(11) I will be here to receive your patrol report when you return.

(12) To communicate with battle group, utilize the AN/PRC-10 with the forward observer. Channel ____. Call sign for the patrol will be ____. Call sign for your battle group FSC, with whom your forward observer will be in contact, is ____.

(13) Challenge is ____. Password is ____.

(14) Weather will be ____. Darkness is at ____. Daylight, ____.

c. *Critique Checklist.*

(1) Did the patrol leader take immediate and positive action when the patrol was subjected to heavy mortar fire (tree charges)?

(2) Did patrol employ stealth to achieve surprise?

(3) Was the action of the patrol leader and members of the patrol logical and tactically sound during seizure of the objective?

(4) Was the reorganization completed in time to repel counterattack?

(5) Did the withdrawal begin as soon as the mission had been accomplished?

(6) Did the patrol leader employ available supporting fires properly?

(7) Was a reconnaissance of the objective made prior to attack?

SUGGESTED TIME SCHEDULE

1100–1145	Briefing.
1145–1200	Issue warning order.
1200–1330	Dinner—prepare for reconnaissance.
1330–1500	Reconnaissance.
1500–1600	Continue planning.
1600–1700	Issue patrol order.
1700–1800	Supper.
1800–1900	Final preparation—rehearsal—inspection.
1900–2000	Move to forward area.
2000–2030	Move to LD.
2030	Cross LD.
2230	Action at objective.
0200–0400	Return through friendly line—return to reserve area—debriefing—critique.

SKETCH OF PROBLEM LAYOUT

Enemy Outpost (Objective)	———————
	1,000 yds
Stream	———————
	1,000 yds
Friendly Company Outpost (L.D.)	———————
	50–200 yds
Friendly Company Command Post	———————
	As desired
Reserve Area	———————

Note. It is recommended that the distance for movement in this exercise be approximately 4,000 yards. However, the distance can be scaled to suit the available terrain. A sufficient number of situations should be included to keep the patrol occupied for a period of five hours after departing the outpost.

Section V. NIGHT RAID AGAINST ENEMY REAR AREA INSTALLATION

14. Purpose

To develop the ability of leaders to prepare a thorough patrol plan and to execute the plan under realistic combat conditions. To familiarize the members of the patrol with the principles and techniques of night patrolling. To develop the ability to infiltrate the enemy's battle position, move in his rear to a designated objective, destroy the objective, and return to friendly lines.

15. Scope

a. Night combat patrol. A night combat patrolling exercise to include planning, preparation, execution, and critique phases. The problem requires the organization of a raid on an enemy rear area position, infiltration of the enemy battle position, movement to and destruction of the objective, and return to friendly lines. See paragraph 3.

b. See FM 21–75.

16. Scenario

a. This problem is a relatively short night combat patrol. It should be scheduled to follow the initial daylight reconnaissance problem and run over the same terrain. It will serve as an introduction to night operations in the area and test the ability of a patrol to infiltrate the Aggressor's main battle position, carry out a combat mission to their rear and return, all in one night. The size of the patrol on this problem should not be more than 30 nor less than 12 men.

b. The patrol is given an orientation. Following the orientation the patrol leader is given an operation order, to include the patrol's specific mission. The patrol leader now becomes responsible for the execution of the proper steps in troop leading procedure required for the successful accomplishment of his mission. Generally, the actions of the patrol leader follows the sequence contained in the word picture of the exercise presented below.

(1) The patrol leader makes an initial estimate of the situation and decides on an efficient way to utilize the time available to him. He makes a thorough map study and comes up with a preliminary plan. He issues a warning order so that preparations can be started by all members. He makes any necessary coordination with personnel available in the area. His coordination completed at battle group level, he and selected subordinates depart for the forward area company through which his patrol is to pass. He coordinates with the company commander and obtains any information that may be available. He then is guided to a forward outpost where he makes a visual reconnaissance. He discusses pertinent details with the individuals in the outpost. The patrol leader and his subordinates now return to the patrol. The patrol leader checks the progress of the

patrol's preparation, and changes his preliminary plan if necessary. He then completes his plan and prepares his patrol order. At the time and place designated in his warning order, he meets with his patrol and issues the patrol order. After answering questions and satisfying himself that his men understand the mission and their respective duties, he dismisses them to complete their preparations. He supervises the patrol and conducts any rehearsals he feels are necessary, such as actions at the objective and at danger areas. When possible, rehearsals are conducted on terrain similar to that over which the patrol is to operate. Prior to departure time the leader conducts a final inspection. At the designated time the patrol is moved forward to the forward company command post area. The patrol leader makes a final check with the company's commanding officer, and then he and his patrol are guided forward to the outpost. The patrol leader checks with the personnel in the outpost for any last minute information they may have. He moves his patrol out and guides them on his selected route. Just outside the friendly position he halts the patrol, establishes security, and listens for signs of a possible enemy ambush. He utilizes pace men, point men, and compass men as desired. He adjusts or changes his formation based on the terrain, cover, concealment, visibility, and proximity to known enemy positions. As the patrol approaches the Aggressor MLR, extreme stealth must be employed if the infiltration is to be made without detection.

(2) The Aggressors manning the line should be alert to exploit any errors committed by the patrol, and should make the crossing of the MLR as difficult as possible. Once discovered, the patrol has several alternatives, and the decisions, actions and orders of the patrol leader during enemy contact provide the basis for a sound evaluation of the patrol leader's ability.

(3) The objectives, whether command posts or mortar positions, must be so located as to be accessible to the patrols, considering the time limits given them and the nature of the terrain over which they must move. Also the locations should be tactically sound. Such a location might include a reverse slope near a road where resupply by vehicle is possible. Aggressor personnel manning these installations must create activity suitable to such objectives. At the regimental command posts vehicles and messengers will be arriving and departing. It is desirable that at least one radio be in operation at the communications center. A fire may be burning and little or no noise discipline will be observed. The mortar positions, in addition to some of these same activities, will be periodically firing missions throughout the night. Shouted fire commands and the sounds of the rounds being fired (simulated by firecrackers) will serve to give the illusion of a mortar platoon in operation. The positions can be built up to in-

clude bunkers, simulated mortars in sandbagged pits, tactical wire, communications wire, and anything else that will add to the realism of the problem.

(4) Once across the MLR, the patrol proceeds to the vicinity of the objective and there puts into effect whatever plan the patrol leader evolves. Approaching the objective, the patrol halts and a reconnaissance is made; last minute changes are put into effect. Actions on the objective must be sound, well planned, and aggressively executed. There must be a plan for withdrawing from the objective, and for reassembling and reorganizing the patrol prior to the return movement back through the MLR.

(5) The same principles that governed the patrol's initial infiltration of the MLR will again apply on the return. If anything, more stealth is required since the entire area is alerted against them once they have completed their mission in the enemy's rear. The patrol leader must again have a plan for action in event of enemy contact and he might consider breaking down into small groups to more easily infiltrate back through the Aggressor MLR.

(6) The patrol must be alert on its return movement through the area between the MLR's. On reaching the appropriate friendly outpost and after going through the proper recognition procedures, the patrol leader counts his patrol back in through the wire, following as the last man in. After checking with the forward company and giving any information

requested, the patrol returns to its reserve area.

(7) On return to the base camp, the patrol is debriefed by the S3. The members then clean their equipment, have breakfast, and rest. After the patrol has rested, it is regrouped and the observer-instructor holds a critique.

17. General and Special Situation

A general and special situation constituting an operation order is presented by the principal instructor who introduces himself as the battle group S3. The following is a suggested method of presenting the situation for this problem.

a. General Situation. Our reconnaissance patrols of the past two days were very successful and brought back much valuable information. It was confirmed that the (unit) Aggressor division holds the ground to our front generally along this line. The strength and weapons employed along this line indicate that this is their MLR. It is believed that their morale is high and that this unit will fight well.

b. Special Situation.

(1) Tonight your company has been directed to send out ____ combat patrols. The S2 has given me the location of ____ Aggressor battalion command posts and ____ mortar positions. These locations were obtained by PW interrogation and by division light aircraft. The pilots were unable to fly low, often coming under heavy small arms fire, so the information is limited. However, the mortar positions are believed to be platoon size.

(2) Your company will send out ____ combat

patrols tonight to attack these installations. Your mission is to kill all the personnel and destroy all the equipment on the objective. The following are your specific objectives: (This is included only if more than one patrol is employed. Coordinates of each objective should be given.)

(3) You will coordinate your departure and return with the forward company commander. You will return through the same friendly positions no earlier than 0430 hours tomorrow. The forward company commanders are prepared to brief you on the situation to their front, and they will provide guides through their lines if you desire.

(The following can be injected into the order if demolitions pits are available on the enemy positions.)

(4) Artillery H & I fires will be falling all along the Aggressor MLR tonight. At ___ hours the artillery liaison officer will be here for coordination. You can arrange with him to shift fires off of your preplanned routes, and also arrange for artillery support both along your route and at the objective. Fires shifted off your routes will remain shifted until all patrols have returned, or unless an emergency develops.

(5) There is a battle group aid station at (coordinates). Patrol leaders will determine disposition of casualties.

(6) You will have a hot meal in your company before leaving and upon return. Battle group transportation will be furnished. Ammuni-

tion and equipment will be furnished by your company.

(7) Radio call signs are ____. The challenge is ____, password is ____.

(8) Here is your time schedule. I will be here on your return to receive your reports. The time is now ____. Are there any questions?

(9) You will cross the LD at ____ hours.

(10) Weather will be ____. Darkness is at ____; daylight, ____.

(11) Are there any questions?

(12) The time is now ____.

 c. *Critique Checklist.*

(1) Was plan for action on objective sound? Were automatic weapons properly employed to isolate objective, cover withdrawal?

(2) Was there a plan for reorganization after action on objective?

(3) Did patrol leader have a plan for employing his automatic weapons en route?

(4) Was isolation of objective during attack accomplished?

SUGGESTED TIME SCHEDULE

0800–0820	Briefing.
0820–0830	Prepare and issue warning order.
0830–0845	Coordination in reserve area.
0845–1015	Travel to forward unit—coordinate—make visual reconnaissance—return to reserve area.
1015–1200	Planning.
1200–1300	Dinner.
1300–1600	Continue planning and preparation.

1600–1700	Issue patrol order.
1700–1800	Supper.
1800–1915	Final preparation, rehearsal, inspection of the patrol.
1915–1945	Travel to forward area.
1945–2015	Movement to LD.
2015	Cross LD.
2015–0430	Accomplish mission—return to friendly lines.
0800	By this time most of the patrols should have returned, been moved back to the reserve area, debriefed, and fed breakfast.
1100	Critique by observer-instructor.

SKETCH OF PROBLEM LAYOUT

Objectives	────────────
	↕ 1,500–2,000 yds
Enemy Main Battle Position	────────────
	↕ 2,000–3,000 yds
Friendly Company Outpost (L.D.)	────────────
	↕ 50–200 yds
Friendly Company Command Post	────────────
	↕ As desired
Reserve Area	────────────

Note. It is recommended that the distance for movement for a patrol in this exercise be approximately 9,000 yards. However, the distance can be scaled to suit the available terrain. A sufficient number of situations should be included to keep the patrol occupied for a period of ten hours after departing the outpost.

Section VI. NIGHT INFILTRATION AND RECONNAISSANCE OF AN AREA DEEP IN ENEMY TERRITORY

18. Purpose

To develop the ability of leaders to prepare a thorough plan and to execute the plan under realistic combat conditions. To familiarize the members of the patrol with the principles and techniques of day and night patrolling. To develop the ability to infiltrate the enemy's battle position and move deep into his rear area. To develop the ability to make a prolonged reconnaissance and surveillance of an area, obtain information, and return to friendly lines.

19. Scope

a. Day and night patrol. A day and night reconnaissance patrolling exercise to include planning, preparation, execution, and critique phases. The problem requires the infiltration of the enemy's battle position, movement in his rear area, organization of an area reconnaissance and surveillance, collection of information, and return to friendly lines. Mission to be accomplished over areas under observation by the enemy, thus necessitating maximum use of stealth, cover, and concealment. See paragraph 3.

b. See FM 21–75.

20. Scenario

a. This exercise requires considerable ability on the part of the patrol leader. It has many situations that require aggressiveness, initiative, and stamina. The patrol will become fatigued and sleepy, thus adding to the problem of the patrol leader. This is a typical ranger type operation. The size of the patrol should not be more than 12 nor less than 6 men.

b. The patrol is given an orientation. Following the orientation, the patrol leader is given an operation order to include the patrol's specific mission. The patrol leader now becomes responsible for the execution of the proper steps in troop leading procedure required for the successful accomplishment of his mission. Generally, the actions of the patrol leader follow the sequence contained in the word picture of the problem presented below.

> (1) The patrol leader makes an initial estimate of the situation and decides on an efficient way to utilize the time available to him. He makes a thorough map study and comes up with a preliminary plan. He issues a warning order so that preparations can be started by all members. He makes any necessary coordination with personnel available in the area. His coordination completed at battle group level, he departs with selected subordinates for the forward area company through which his patrol is to pass. He coordinates with the company commander and obtains all information that may be available. He then is guided to a forward outpost where he makes a visual reconnaissance of the initial route to the objective. He discusses pertinent details with the individuals in the outpost. The patrol leader now returns to his patrol; he checks the progress of their preparation and changes his preliminary plan if necessary. He then completes his plan and prepares his patrol order. At the

time and place designated in the warning order, he meets with his patrol and issues the patrol order. After answering questions and satisfying himself that his men understand the mission and their respective duties, he dismisses them to complete their preparations. He supervises the patrol and conducts any rehearsals he feels are necessary, such as actions at danger areas and at the objective. When possible, rehearsals are conducted on terrain similar to that over which the patrol is to operate. Prior to departure, the leader conducts a final inspection. At the designated time, the patrol is moved to the forward company command post area. The patrol leader makes a final check with the company commanding officer, and then he and his patrol are guided forward to the outpost. The patrol leader checks with the personnel in the outpost for any last minute information they may have. He moves the patrol out and guides them on the selected route. He utilizes pace men, point men, and compass men as desired. He adjusts or changes his formation based upon the terrain, cover, concealment, visibility, and proximity to known enemy positions.

(2) Approximately 100 yards in front of the outpost there is a small Aggressor party that fires at the patrol and then pulls out of the area. This forces the patrol leader into a course of action. The observer may desire at this time to have the patrol leader become a casualty.

(3) Once the patrol is reformed, it moves to the enemy main line, where it infiltrates during the time that friendly artillery shifts to the right or left.

(4) Once behind the main line, it moves over difficult terrain to an area where it sets up a clandestine bivouac. It should arrive at this area before daylight. The bivouac should offer concealment and security, be easy to defend, and afford observation of the objective area. It may be possible to locate it near a stream so water is available to the patrol.

(5) At this bivouac area the men must set up security and have the reconnaissance elements rechecked on their jobs. As the objective is an area target, there should be several small reconnaissance elements assigned to various portions. The plan should include time of return to the patrol bivouac area. An alternate clandestine area is picked in case the first becomes known to the Aggressor. The reconnaissance elements accomplish their missions during the daylight hours. Upon their return, all information is disseminated to the entire patrol.

(6) If during the conduct of the action at the objective any member of the patrol becomes captured, he should try to escape as soon as possible. Security by the Aggressors, after questioning and interrogation of the prisoner, should be relaxed, thereby enhancing escape and allowing the prisoner time to rejoin his patrol prior to its departure on the return

route. Aggressor forces, just prior to darkness the second night, will then move to ambush sites on logical routes of movement that the patrol might take in its return. The action of the Aggressor here is logical since the prisoner indicated the likely presence of an enemy force in the area.

(7) The patrol must return through the Aggressor MLR during that time in which the artillery has been shifted. Prior to the patrol's return through the Aggressor MLR, Aggressor troops are moved from the ambush locations for the purpose of reestablishing their MLR. The patrol is told that they may not enter their friendly lines until around daylight hours, thus allowing more time for the reestablishment of the friendly lines' representative group.

(8) The patrol passes through the friendly lines after having been properly challenged, are moved back to the reserve area, debriefed by the S3, and fed breakfast. The observer-instructor then conducts the critique.

21. General and Special Situation

A general and special situation constituting an operation order is presented by the principal instructor who introduces himself as the battle group S3. The following is a suggested method of presenting the situation for the problem:

a. General Situation.

(1) Enemy positions are as shown on this situation map. There have been no changes in the past 72 hours. From our past engagements

with Aggressor, we have found him firmly entrenched and very aggressive. He has excellent mortar and artillery support. Our latest intelligence indicates that he is busy preparing many new installations several miles to his rear.

    (2) Our situation has remained the same for the past five days. It is as shown on this situation map.

b. *Special Situation.*
    (1) You have been selected to lead a reconnaissance patrol deep into the enemy rear area.
    (2) Your mission is to make a reconnaissance of the area as indicated on this overlay.
    (3) Your general route is indicated by several check points on this overlay. No other patrols will be operating in your general area.
    (4) Leave this area at ____ hours and pass through ____ company of the ____ battle group at ____ hours. Return through the same unit by ____ hours on (date). Not before.
    (5) Artillery H & I fires on the Aggressor positions will be shifted to the left or right from ____ to ____ hours tonight and ____ to ____ hours on (date) so you can safely infiltrate the Aggressor position. They will continue to fall on both of your flanks. Supporting fires will be available to you at the main Aggressor battle position only.
    (6) Battle group transportation will be furnished.
    (7) Turn request for all supplies in immediately after your warning order. Pick them up at ____ hours.

(8) Route of evacuation of casualties will be back through ____ company. If practical, take prisoners.
(9) Challenge and password (for 1st, 2d, 3d day).
(10) Weather will be ____. Darkness is at ____, daylight, ____.
(11) Are there any questions?
(12) Time is now ____ hours.

c. *Critique Checklist.*
(1) Discuss infiltration of enemy MLR.
(2) Was action of patrol on contact with Aggressor tactically sound?
(3) Did patrol leader inspect his patrol during rest breaks for faulty weapons and condition of patrol members?
(4) Did patrol leader make provisions for reorganization after accomplishment of mission?
(5) Did the reconnaissance element of patrol disseminate information to security element?
(6) Were passwords properly used?
(7) Discuss patrol base and clandestine bivouac area selected.
(8) Did patrol practice radio security?
(9) Was action at objective tactically sound?

SUGGESTED TIME SCHEDULE

0800–0900	Briefing.
0900–0915	Warning order issued.
0915–1200	Planning and coordination.
1200–1300	Dinner.
1300–1430	Continue Planning.
1430–1600	Move to forward company area—coordination—reconnaissance return.

1600–1700	Issue patrol order.
1700–1800	Supper.
1800–1900	Final preparation—rehearsal—inspection.
1900–1930	Move to forward area.
1930–2000	Move to LD.
2000	Cross LD.
2d day	
0600	Arrive at clandestine bivouac.
0600–1800	Reconnaissance of objective.
1800	Depart clandestine bivouac.
3d day	
0400	Arrive friendly lines.
0430	Arrive reserve area.
0430–0600	Debriefing—breakfast—critique.

SKETCH OF PROBLEM LAYOUT

Objective Area ─────────
                          ▲
                          5–7 miles
                          ▼
Enemy MLR   ─────────
                          ▲
                          1,800–2,000 yds
                          ▼
Friendly Forward Company Outpost (L.D.) ─────────
                          ▲
                          100–200 yds
                          ▼
Forward Company CP ─────────
                          ▲
                          As desired
                          ▼
Reserve Area ─────────

Note. It is recommended that the distance for movement for a patrol in this exercise be approximately 14 miles. However, the distance can be scaled to suit the available terrain. A sufficient number of situations should be included to keep the patrol occupied for a period of 36 hours after departing the outpost.

Section VII. RAID AGAINST ENEMY GUERILLA CAMP

22. Purpose

To develop the ability to prepare a thorough patrol plan and to execute the plan under realistic combat conditions. To familiarize the members of the patrol with the principles and techniques of patrolling. To develop the ability to move close to an enemy guerilla installation, avoid his security, and raid the guerilla camp.

23. Scope

a. Day combat patrol exercise. A day combat patrolling exercise to include planning, preparation, execution, and critique phases. The problem requires the evaluation of information concerning an enemy guerilla installation, organization of a raid on the installation, movement by stealth to the area of the guerilla camp, conduct of the raid, destruction of all personnel and equipment, and return to patrol base. See paragraph 3.

b. See FM 7-10 and FM 21-75.

24. Scenario

a. This exercise has many situations that require aggressiveness, initiative, and stamina. The size of the patrol should not be more than 40 nor less than 18 men.

b. Following the orientation, the patrol leader is given an operation order to include the patrol's specific mission. The patrol leader now becomes responsible for the execution of the proper steps in troop leading procedure required for the successful accomplishment of his mission. Generally, the actions of the patrol leader follow the sequence contained in the word picture of the problem presented below.

(1) The patrol leader makes an initial estimate of the situation and decides on an efficient way to utilize the time available to him. He makes a thorough map study and comes up with a preliminary plan. He issues a warning order so that preparations can be started by all members. He makes any necessary coordination with personnel available in the area. His coordination completed, he checks the progress of the patrol's preparation. He makes changes in his preliminary plan if necessary. He then completes his plan and prepares his patrol order. At the time and place designated in his warning order, he meets with his patrol and issues the patrol order. After answering questions and satisfying himself that his men understand the mission and their respective duties, he dismisses them to complete their preparations. He supervises the patrol and conducts any rehearsals that he feels are necessary. When possible, rehearsals are conducted on terrain similar to that over which the patrol is to operate. Prior to departure time, the patrol leader conducts a final inspection. At the designated time, the patrol is moved to a patrol base command post area. This area should appear to be occupied by at least a platoon of men. It should give the appearance of a perimeter defense since it is located in rear of friendly lines in a guerilla infested area. The command post area should consist of at least two foxholes or one bunker. The patrol leader makes a final check with

the officer in charge prior to the patrol's departure from the area.

(2) The objective area should be located near a stream or road that would logically serve as a means of withdrawal or resupply for the camp. The area around the guerilla camp should be heavily booby trapped. In addition, there should be numerous warning devices to alert Aggressor personnel of the patrol's approach. Aggressor lookouts should be placed around the camp to warn the camp of the arrival of the patrol. Activity within the camp should be logical and realistic.

(3) The vehicles carrying the patrol to the patrol base should be moved tactically. On the way to the patrol base, the patrol is ambushed by Aggressors. The site for the ambush should be logically chosen with demolition pits on one side of the road and Aggressor troops with many automatic weapons and grenades on the other side. After taking action against the ambush, the patrol leader has the convoy continue on the patrol base. There, last minute coordination is accomplished and the patrol is resupplied with ammunition, if it is requested. After the patrol leaves the patrol base, it is fired upon by isolated guerilla security posts. The patrol leader takes necessary action and the patrol continues to the objective.

(4) As the patrol nears the objective and is detected, the guards alert the guerillas. The guerillas set up a hasty defense and commence a withdrawal. At the designated time, or

upon detection, the patrol leader puts his plan into effect. Once on the objective, prisoners are taken, casualties searched, and the installation destroyed. Upon completion of the mission, the patrol withdraws. During the return, a small group of guerillas ambush the patrol a few hundred yards from the patrol base. After taking necessary action, the patrol returns to the base and moves to the reserve area. There it is debriefed by the battle group S2 or S3. The patrol is then critiqued by the observer-instructor.

25. General and Special Situation

A general and special situation constituting an operation order is presented by the principal instructor who introduces himself as the S2 or S3, as desired. The following is a suggested method of presenting the situation for this problem:

a. General Situation.
   (1) Enemy are as shown on this overlay. Notice these guerillas here (indicate on overlay) behind our lines. During our last big attack, we bypassed several small isolated units that made their way into the area here and have been conducting guerilla-type activities along our main supply route. Last night we had a supply convoy ambushed here (point). The (unit) had another one ambushed here (indicate) the night before. Our intelligence indicates that their camp is located along this (stream, road).
   (2) Our lines are as shown here (indicate on overlay). No changes in disposition. We have

established a patrol base here (point) in the center of the guerilla territory. This patrol base has been sending reconnaissance patrols into the area.

b. *Special Situation.*
   (1) Division has received information through reconnaissance patrols that the guerilla headquarters is located at coordinates ____. We have been ordered to kill or capture these guerillas and destroy their camp. You have been selected to do this job.
   (2) There will be no fire support.
   (3) There will be no other patrols operating in this area.
   (4) Your mission is to destroy this camp, kill or capture all personnel, and destroy their equipment.
   (5) You will leave this area at ____ hours by vehicle to our patrol base located at coordinates ____. You will leave the patrol base at ____ hours and return upon completion of your mission.
   (6) Remember, you are fighting guerillas. Be careful of an ambush or security guards as you approach your objective. Don't let them escape.
   (7) You will report by radio as soon as your mission is accomplished. You will be debriefed by ____ upon your return.
   (8) Request for ammunition, equipment, and necessary rations will be turned in by ____ hours.

(9) Battle group transportation will be furnished to the patrol base. It will be located at ____ at (time) hours.

(10) Your call sign is ____, your frequency is ____. battle group will be ____.

(11) Challenge and password: ____, ____.

(12) Weather will be ____. Darkness is at ____, daylight, ____.

(13) Are there any questions?

(14) Time is now ____ hours.

c. *Critique Checklist.*
   (1) Discuss actions and orders of patrol leader when vehicle is ambushed.
   (2) Did patrol leader coordinate properly with the commander of the patrol base?
   (3) Was stealth used to achieve surprise?
   (4) Discuss action at objective. Were measures taken to cut off escape before attack started? Were trees checked for snipers? Was camp thoroughly destroyed? Were prisoners taken? Was an aggressive assault conducted? Did patrol maintain proper security after the assault?

SUGGESTED TIME SCHEDULE

2200–2230	Briefing.
2230–0230	Preparation—issue warning order—continue planning.
0230–0315	Issue patrol order.
0315–0445	Rehearsal—final preparation.
0445–0530	Breakfast.
0530–0600	Move to patrol base.

0600–0630	Coordination at patrol base.
0630	Cross LD.
1200–1230	Action at objective.
1700	Arrive at patrol base.
1700–2000	Return reserve area—debriefed—supper—critiqued.

SKETCH OF PROBLEM LAYOUT

Objective ─────────────
  ▲
  3 miles
  ▼
Patrol Base (L.D.) ─────────────
  ▲
  As desired
  ▼
Ambush Site ─────────────
  ▲
  As desired
  ▼
Reserve Area ─────────────

Note. It is recommended that the distance for movement for a patrol in this exercise be approximately 6 miles. However, the distance can be scaled to suit the available terrain. A sufficient number of situations should be included to keep the patrol occupied for a period of ten hours after departing the patrol base.

Section VIII. RAID TO SEIZE AND HOLD KEY ENEMY INSTALLATION

26. Purpose

To develop the ability of leaders to prepare a thorough patrol plan and to execute the plan under realistic combat conditions. To familiarize the members of the patrol with the principles and techniques of day and night patrolling and perimeter defense. To develop the ability to infiltrate the enemy's battle position, seize and hold an objective, and at the same time prevent the destruction of the objective installation by the retreating enemy.

27. Scope

a. Day and night combat patrol. A day and night combat patrolling exercise to include planning, preparation, execution, and critique phases. The problem requires infiltrating the enemy's battle position, moving three to five miles to an objective in enemy territory, seizing and defending the objective, receiving resupply and medical evacuation by air, and executing a night relief on position by friendly forces. See paragraph 3.

b. See FM 7-10 and FM 21-75.

28. Scenario

a. Before participating in this exercise, troops should have a thorough knowledge of map reading, use of compass, hand to hand combat, night patrolling, infiltration, conduct of the defense, aerial resupply, and evacuation of wounded by helicopter. Troops should have participated in at least one night combat patrol and one night reconnaissance patrol involving infiltration of enemy battle positions. The objective area is fairly isolated from enemy supply routes and is on high ground suitable for a perimeter defense by a patrol of 20 to 46 men. Near the objective there should be a suitable landing site and drop zone for helicopter evacuation of wounded and aerial resupply by parachute or free fall. The objective is a simulated radar station or some other type installation. A barbed wire fence should be placed around the installation. There should be a gate and guard shack at the only entrance. It is desired that demolition pits be placed around the objective and connected to a demolition board at a control point. The charges in these pits represent friendly artillery support and are fired on call from the patrol leader. The control point is constructed in a concealed position approximately 100 yards from the objective.

b. The patrol is given an orientation. The patrol leader then receives an operation order to include his patrol's specific mission. Maps, overlays, aerial photographs, and terrain models may be used in conjunction with the issuance of this operation order; however, the types and quantities of the visual aids employed should be consistent with those normally found under similar circumstances in a combat situation.

c. The patrol leader now becomes responsible for the execution of the proper steps in troop leading procedure required for the successful accomplishment of his mission. Generally, the actions of the patrol leader follow the sequence contained in the word picture of the problem presented below.

> (1) The patrol leader makes an initial estimate of the situation and decides on an efficient way to utilize the time with a preliminary plan. He issues a warning order so that preparation can be started by all members. He makes any necessary coordination with personnel available in the area. His coordination at battle group level completed, he departs for the forward company through which his patrol is to pass. He coordinates with the company commander and obtains any information that may be available. Then he is guided to a forward outpost where he makes a visual reconnaissance of the terrain between his lines and the enemy's lines. He discusses pertinent details with the individuals in the outpost. The patrol leader then returns to his patrol and checks the progress of prepara-

tion. He makes changes in his preliminary plan if necessary. Then he completes his plan and prepares his patrol order. At the time and place designated in his warning order, he meets his patrol and issues the patrol order. After answering questions and satisfying himself that his men understand the mission and their duties, he dismisses them to complete their preparations. He supervises the patrol and conducts any rehearsals that he feels are necessary, such as actions at the objective and at danger areas. When possible, rehearsals are conducted on terrain similar to that over which the patrol is to operate. Before departure time, the patrol leader conducts a final inspection.

(2) At the designated time, the patrol is moved to the forward company command post area. The patrol is met by a guide from this company who takes the patrol to the company area. As the patrol nears the company command post, it is hit by an Aggressor artillery barrage. This barrage is caused by simulator shell, ground burst, and/or demolition pits. In the midst of the artillery barrage, simulated casualties, played by control personnel, yell and scream for the medic. Litter teams pick up wounded and evacuate them. This action should be made as realistic as possible. The guide points out the command post to the patrol leader and disappears.

(3) The patrol leader is then faced with the problems of properly dispersing his men, maintaining control, caring for casualties, and

making final coordination. The company commander briefs the patrol leader on the latest developments. He points out on his map the location of a known gap in the enemy MLR and tells the patrol leader that he will furnish a guide to this gap. The guide reappears at the command post and guides the patrol through the friendly MLR. The patrol is challenged as it moves through the MLR and is allowed to pass. It continues to the friendly outpost where it is again challenged. The patrol leader is recognized by the outpost personnel and told to move on. The guide then leads the patrol forward until it is approximately 100 yards from the gap. (The guide then leaves the patrol and returns to friendly lines.)

(4) The patrol begins the infiltration of the enemy battle position. A reconnaissance team should be sent forward to verify the location of the reported gap. If the patrol is able to infiltrate without being detected, there is no Aggressor action at the MLR. However, if the patrol is detected, the Aggressor control officer observes the infiltration of the patrol until one-half of the patrol has passed through the gap, then the signal is given for the Aggressors to open fire. Flares illuminate the area, and the patrol receives heavy rifle fire, machine gun fire, and simulator grenades. Selected members of the patrol are declared casualties by the observer-instructor. The patrol leader is told that the casualties have been hit in the stomach and in the legs and

cannot move. The patrol leader decides whether to leave them or to evacuate them. Enemy firing continues until the patrol leader either fights his way through the MLR or withdraws out of the area. The Aggressor control officer determines the time to cease fire.

(5) The Aggressors capture as many of the friendly patrol as possible. Then they are moved to the rear of the enemy position by guards who have been instructed to use lax security measures. The captured patrol members should escape during their movement to the rear and rejoin the patrol at a designated rallying point near the objective.

(6) The patrol leader reorganizes his patrol and moves cross-country to the objective area, avoiding roads, trails, and likely locations of Aggressor rear area installations. During the movement to the objective, Aggressor motorized patrols are active along all usable roads. At prominent road junctions and bridges, there are Aggressor security guards or traffic control teams. If the patrol is detected, it is challenged in the Aggressor language. When no password is forthcoming, the Aggressors open fire and report the location of the patrol to the Aggressor control officer. If the patrol attacks the Aggressors, they hastily retreat and hide until the friendly patrol departs. The Aggressor control officer, using all information reported to him, takes necessary action to locate and capture the friendly patrol. Members of the patrol

who are captured are moved to an Aggressor headquarters area where they are interrogated, threatened, and loosely guarded. The patrol members should attempt to escape and rejoin the patrol.

(7) The patrol should arrive in the vicinity of the objective two to four hours before dawn. It should occupy a clandestine bivouac area while the patrol leader makes a detailed reconnaissance of the objective. About one hour before dawn, the patrol leader should move the patrol close to the objective. Killer teams should be sent out to eliminate the walking guards around the objective and the Aggressor guard detachment. Should the killer teams be detected, a base of fire and an assault element should be in position and ready to start an attack. When the position has been captured, a hasty perimeter is established, Aggressor prisoners are secured, ammunition is redistributed, and wounded are treated. Aggressors leave several automatic weapons and a specified quantity of ammunition and equipment to be captured.

(8) The Aggressors counterattack thirty minutes after they withdraw. They advance to within 20 yards of the perimeter, engage the patrol, then fall back in disorder to their control point. During the remainder of the problem, the Aggressor control officer should conduct several attacks against the perimeter, withdrawing after each encounter. He may employ snipers instead of the unit attacks.

(9) Four hours after dawn, an aerial resupply by

parachute or free fall is made by division light aircraft. The patrol leader establishes security and prepares the drop zone. There is no Aggressor action during the resupply unless the bundles are dropped outside the drop zone. If the patrol leader sends recovery parties out away from supporting fires of the perimeter to recover the bundles, Aggressors will attempt to capture patrol members and their supplies. The resupply will consist of ammunition and weapons, but will not contain food or water. If bad weather prohibits an aerial resupply, the patrol leader is informed. He is required to continue the mission.

(10) At 1400 hours, the observer-instructor declares designated casualties to be in critical condition. The patrol leader should contact battle group S3 by radio and request helicopter evacuation. Weather permitting, helicopters are sent at a prearranged time to evacuate casualties. There is no Aggressor action during the helicopter evacuation.

(11) If men become desperate for water and food, they are required to forage for it. Aggressors, where possible, capture foraging parties.

(12) During the day the patrol should constantly improve their positions, install warning devices, listening posts, improve fire lanes, and construct overhead cover. The patrol should use all available cover and concealment throughout the defense.

(13) All Aggressor movement near the perimeter must utilize cover and concealment. There should not be a continuous harassing of the

perimeter throughout the defensive phase. The patrol should be lulled into a false sense of security by long periods of inactivity, and then shocked by violent attacks.

(14) During the briefing, the patrol was told that it would be relieved one hour after dusk. At dusk the S2 informs the patrol leader by radio that relief is delayed and that the patrol is to remain on position until relieved. At 2300 hours, the S2 tells the patrol leader that he will be relieved by elements of the ____ battle group at 2000 hours, and will move to a location one mile from the perimeter to meet transportation.

(15) After this is accomplished, battle group transportation moves the patrol to the reserve area. The patrol is debriefed by the S2. Patrol members clean their equipment, rest, and are critiqued by the observer-instructor.

29. General and Special Situation

A general and special situation constituting an operation order is presented by the principal instructor who introduces himself as the S2 or S3 as desired. The following is a suggested method of presenting the situation of this problem.

  a. *General Situation.*
  (1) Enemy positions are as shown on this overlay. Aggressor morale is believed to be low.
  (2) The division attacks at 0500 hours tomorrow morning, launching a general offensive.

  b. *Special Situation.*
  (1) Division has received from line crossers and light aircraft observers the location of a key

(installation) deep in the enemy's rear area. We have been ordered to seize and hold this installation before the general attack commences tomorrow morning.

(2) Your patrol will infiltrate the enemy MLR and proceed to this installation, seize it before 0500 tomorrow, and hold it until relieved by friendly forces.

(3) Relief will be effected by 2000 hours tomorrow night by elements of the ____ division.

(4) Do not destroy the installation unless forced off position. It is wanted for study by our intelligence experts.

(5) Move through ____ company of the ____ battle group at ____ hours.

(6) A reconnaissance patrol has reported a gap in the Aggressor wire located at coordinates _____. Prearranged artillery fires will fall on both sides of this gap in order to cover your movement.

(7) A guide will meet you at the detrucking point and take you to the company command post and to the outpost.

(8) Artillery support is available throughout the operation. Turn in request for preplanned fires at ____ hours.

(9) No other friendly patrols are operating in your area.

(10) Request for additional weapons, ammunition, and equipment will be turned in at ____ hours. No rations will be taken on the patrol. Arrangements will be made for a

resupply by air at 0800 hours tomorrow. Turn in a separate request for supplies to be air dropped.

(11) Battle group transportation to the forward area will be furnished. Be prepared to leave reserve area by ____ hours.

(12) Helicopters are available for evacuation of casualties during daylight hours. Request them as required.

(13) Radio call signs and frequencies:
Your patrol will be ____ operating on frequency ____.
Battle group S3 is ____.

(14) Challenge ____ (For each day of the problem).
Password ____ (For each day of the problem).

(15) Weather will be ____. Darkness is at ____. Daylight, ____.

c. *Critique Checklist.*
  (1) Was infiltration of enemy MLR successful? Comment on patrol leader's action.
  (2) Did patrol use stealth to achieve surprise?
  (3) Were security posts eliminated silently?
  (4) Cover action of patrol leader, squad leaders, and members of patrol during the seizing of the objective.
  (5) Was organization of objective completed in time to repel counterattack?
  (6) Did patrol leader improve the disposition of weapons and personnel with the coming of dawn?
  (7) Were defensive fortifications constructed and improved throughout defensive phase?

(8) Did patrol leader properly utilize supporting artillery fire?

(9) Critique actions of patrol leader and patrol members during each Aggressor action against the perimeter.

(10) Was ammunition redistributed after each attack?

(11) Discuss aerial resupply and helicopter evacuation of wounded.

(12) Was relieving platoon properly challenged and received by patrol? Comment on night relief.

SUGGESTED TIME SCHEDULE

0800–0900	Briefing.
0900–0915	Issue warning order.
0915–1200	Coordination—preparation.
1200–1300	Dinner.
1300–1430	Continue planning.
1430–1630	Move to forward area—coordination—visual reconnaissance—move back to reserve area.
1630–1730	Issue patrol order.
1730–1830	Supper.
1830–1930	Final preparation—rehearsal—inspection.
1930–2000	Move to forward area.
2000–2030	Move to LD.
2030	Cross to LD.
0600	Arrive objective.
0600–2000	Occupy objective.
2000	Relief on objective.
2000–2400	Move to reserve area—debriefing—critique.

SKETCH OF PROBLEM LAYOUT

Objective Installation	────────────
	▲ 3–5 miles ▼
Enemy Main Battle Position	────────────
	▲ 1,800–2,000 yds ▼
Forward Company Outpost (L.D.)	────────────
	▲ 50–200 yds ▼
Forward Company Command Post	────────────
	▲ As desired ▼
Reserve Area	────────────

Note. It is recommended that the distance for movement for a patrol in this exercise be approximately six miles. However, the distance can be scaled to suit the available terrain. A sufficient number of situations should be included to keep the patrol occupied for a period of 24 hours after departing the outpost.

Section IX. NIGHT INFILTRATION AND AMBUSH

30. Purpose

To develop the ability of leaders to prepare a thorough patrol plan and to execute that plan under realistic combat conditions. To familiarize the members of the patrol with the principles and techniques of night patrolling. To develop the ability to infiltrate the enemy's battle position and move in his rear area to a designated supply route. To develop the ability to ambush the enemy and establish a roadblock, withdraw, and make contact with a friendly unit.

31. Scope

a. Night combat patrol. A night combat patrolling exercise to include planning, preparation, execution, and critique phases. The problem requires the infiltration of the enemy's battle position, movement to an area designated for an ambush, selection of the ambush site, ambush of a vehicular convoy, destruction of the convoy, the blocking of the supply route with the destroyed vehicles, and movement to a designated rendezvous point to contact friendly forces. See paragraph 3.

b. See FM 7–10 and FM 21–75.

32. Scenario

a. This exercise should follow a conference on ambushes and roadblocks as covered in the preparatory phase. This problem requires the application of the principles obtained in the conference. In addition, the application of the principles of night patrolling and infiltration is required. The only particular feature required for the conduct of the problem is a suitable road in the enemy's rear that may serve as one of his main supply routes. Additional personnel and vehicles are required to form the Aggressor convoy that is to be ambushed. A *minimum* number of required vehicles is three 2½-ton trucks and one ¼-ton truck. These vehicles should be open and the troops should be evenly distributed throughout the convoy. The size of the patrol should be 16 to 36 men.

b. The patrol is oriented. Following the orientation, the patrol leader receives an operation order. The patrol leader now becomes responsible for the execution of the proper steps in troop leading procedure required for the successful accomplishment of his mission.

c. Generally, the actions of the patrol leader follow the sequence contained in the word picture given below.

(1) The patrol leader makes an initial estimate of the situation and decides on an efficient way to utilize the time available to him. He makes a thorough map study and forms a preliminary plan. He issues a warning order so that preparation can be started by all members. He coordinates with personnel available in the area. His coordination completed at battle group level, he departs for the forward company area through which his patrol is to pass. He coordinates with the company commander and obtains any available information. Then he is guided to a forward outpost where he makes a visual reconnaissance of the terrain. He discusses pertinent details with the individuals in the outpost. The patrol leader returns to his patrol and checks the progress of their preparation. He makes changes to his preliminary plan if necessary. He then completes his plan and prepares his patrol order. At the time and place designated in his warning order, he meets his patrol and issues the patrol order. After answering questions and satisfying himself that his men understand the mission and their duties, he dismisses them to complete their preparations. He supervises the patrol and conducts any rehearsals he feels are necessary. When possible, rehearsals are conducted on terrain similar to that over which the patrol is to operate. Prior to departure time, the leader conducts a final inspection. At the designated time, the patrol is moved forward to the forward company

command post area. The patrol leader makes a final check with the forward company commander and then he and his patrol are guided forward to the outpost. The patrol leader checks with the personnel in the outpost for any information they may have. He moves his patrol out and guides it on to his selected route. He utilizes pace men, point men, and compass men as desired. He changes his formation as required by terrain, cover, concealment, visibility, and proximity to known enemy positions. The patrol halts just outside of friendly lines, establishes security, and listens for signs of enemy ambush. Satisfied that there is none, the patrol continues. As the patrol nears the enemy lines, the patrol leader should slow his rate of march, select his route of approach, and enforce noise discipline. If the patrol moves silently along a good route, it may be successful in infiltrating the enemy's line without being detected. In the initial briefing, known enemy positions are pointed out and the patrol leader avoids those areas. If the patrol is detected by the Aggressor, it will be allowed to infiltrate until it is close to the position, then the Aggressor opens fire. The Aggressor remains in position until the patrol pushes on or withdraws.

(2) When the patrol is through the enemy MLR, it proceeds to the ambush site. The purpose of the ambush-roadblock is to delay the enemy in his withdrawal and to hinder the resupply of his front lines. Maximum boundaries limiting the location of the ambush site

are ordered by the battle group S3 in his operation order; however, the patrol leader must choose the exact site. Aggressor vehicles should be utilized on the supply route in the following manner: single vehicles should traverse the road periodically in order to tempt the patrol leader to commit his patrol; Aggressor convoys should traverse the road to present the ambush target to the patrol several times during the hours allotted for the ambush in the operation order.

(3) The patrol leader, utilizing surprise, opens fire on the convoy in heavy volume. Supporting fires are utilized to isolate the convoy and to contain the personnel in the ambush site. The dead are searched and the vehicles are set on fire. The patrol then moves to a preselected assembly point. This location has been previously reconnoitered. Reorganization, if necessary, takes place here. The patrol then moves from the area as rapidly as possible to avoid countermeasures from any Aggressors that might have been alerted by the fire fight. It continues to a rendezvous point where it contacts a friendly patrol for the return to the reserve area. This contact patrol is part of the friendly force that has attacked the enemy positions during the night, and thus can be logically placed in the exercise. The plans for this attack are revealed to the patrol in the operation order.

(4) After returning, the patrol is immediately debriefed by the S2 and fed breakfast. After

cleaning its equipment and resting, it is critiqued by the observer-instructor.

33. General and Special Situation

A general and special situation constituting an operation order is presented by the principal instructor who introduces himself as the S2 or S3. The following is a suggested method of presenting the situation for this problem:

a. General Situation.
  (1) The enemy situation remains about the same. He continues to occupy positions as shown on the overlay.
  (2) Our division attacks at 0030 hours tomorrow as part of a general offensive.

b. Special Situation.
  (1) Your (platoon, squad) has been selected to conduct an ambush behind enemy lines and establish a roadblock to delay the enemy's withdrawal and prevent him from resupplying his lines during our attack.
  (2) Your mission is to establish a roadblock by means of an ambush of a worthwhile vehicular target along the road between coordinates ____ and ____. This ambush-roadblock is to be established between ____ and ____ hours tomorrow morning. (The time should be the most logical times in which the road would be used for withdrawal and resupply resulting from the attack at 0030 hrs.)
  (3) The ambush will not be established nor the road blocked prior to 0230. (This allows control time in which to get the Aggressor

personnel manning the MLR out of position and into the convoy trucks, and also the time required to get the convoys operating on the road.)

(4) After the mission is accomplished, or if at 0430 it is not accomplished, you will move to coordinates _ _ _ _, establish security, and rendezvous with a friendly contact patrol. If contact is not made by 1000 hours, withdraw to coordinates _ _ _ _ _. (This should be the location of the reserve area. Of course, contact *should* be made. This statement is included in the order to suggest action in case the attack is a failure and to thus maintain realism.)

(5) Depart this area at _ _ _ _ hours.

(6) Pass through _ _ _ _ company of the _ _ _ _ battle group at coordinates _ _ _ .

(7) Cross the LD at _ _ _ _ hours.

(8) Ammunition and equipment will be requested and drawn through your company.

(9) Patrol leader will determine the disposition of casualties and prisoners.

(10) Challenge: _ _ _ _ (For each day of problem)
Password: _ _ _ _ (For each day of problem)

(11) Weather will be _ _ _ _ _. Darkness is at _ _ _ _, daylight, _ _ _ _ _.

(12) Any questions?

c. *Critique Checklist.*

(1) Was patrol broken into small teams for the infiltration of enemy battle position? Was method tactically sound?

(2) Was the ambush site tactically sound?

(3) Was surprise achieved?

(4) Discuss organization of ambush site.

(5) Discuss action during ambush.

(6) Discuss withdrawal from ambush site and movement to rendezvous point.

SUGGESTED TIME SCHEDULE

0800–0900	Briefing.
0900–0915	Prepare and issue warning order.
0915–1200	Coordination—continue planning.
1200–1300	Dinner.
1300–1430	Continue planning.
1430–1600	Move to forward area—coordination—reconnaissance—return.
1600–1700	Issue patrol order.
1700–1800	Supper.
1800–1930	Final preparation—rehearsal—inspection.
1930–2000	Move to forward area.
2000–2030	Move to LD.
2030	Cross LD.
0230–0430	Ambush.
0700	Contact friendly forces.
0800	Return to reserve area—debriefing—critique.

SKETCH OF PROBLEM LAYOUT

Rendezvous Point	────────────
	↕ 1,000–1,500 yds
Ambush Site	────────────
	↕ 1,000–1,500 yds
Enemy MLR	────────────
	↕ 1,800–2,000 yds
Outpost (L.D.)	────────────
	↕ 100–200 yds
Forward Company Command Post Area	────────────
	↕ As desired
Reserve Area	────────────

Note. It is recommended that the distance for movement for a patrol in this exercise be approximately 4,400 yards. However, the distance can be scaled to suit the available terrain. A sufficient number of situations should be included to keep the patrol occupied for a period of eleven hours after departing the outpost.

Section X. WATERBORNE RAID AGAINST CRITICAL INSTALLATION

34. Purpose

To develop the ability of leaders to prepare a thorough patrol plan and to execute the plan under realistic combat conditions. To familiarize the members of the patrol with the principles and techniques of patrolling. To develop the ability to use small boats for assault landings. To develop the ability to move in the enemy's rear areas to a designated installation, destroy the installation, and take prisoners.

35. Scope

a. This exercise is adaptable only to those training areas in which a coast is available. It is generally a waterborne raid and requires coordination with naval operations, in addition to the required special terrain called for above. The exercise will have to be revised if it is desired that it be employed in those areas lacking these particular requirements.

b. A day and night combat patrolling exercise to include planning, preparation, execution, and critique phases. The problem to require coordination with friendly units, selection of embarkation and debarkation sites for small boats, organization of boat teams, security while waterborne, night landing, safeguarding of boats, movement through enemy rear areas, destruction of an objective, release of prisoners, movement to clandestine bivouac area, withdrawal through enemy rear areas, and rendezvous with craft off shore. See paragraph 3.

c. See FM 7–10 and FM 21–75.

36. Scenario

a. This exercise should be conducted during the final stages of training because it repeats most of the principles taught in earlier training. It contains many situations and requires numerous decisions by the patrol leaders. It thoroughly tests the patrol leader, his chain of command, every individual of the patrol, and the patrol as a team. The size of the patrol on the problem should not be more than 40 nor less than 18 men.

b. The patrol is oriented and the patrol leader is presented with the operation order to include his patrol's specific mission. This is accomplished on the battle group level. The patrol leader now becomes responsible for the execution of the proper steps in troop leading procedure required for the successful accomplishment of his mission. Generally, the actions of the patrol leader follow the sequence outlined in the word picture presented below.

    (1) The patrol leader makes an initial estimate of the situation and decides on an efficient way to utilize the time available to him. He makes a thorough map study and forms a preliminary plan. He issues a warning order so that preparation can be started by all members. He coordinates with personnel available in the area, to include the naval officer involved in the exercise. The naval officer should assist the battle group S3 in his orientation, to include briefing the patrol leader on details pertinent to the naval support of the exercise. When his coordination has been completed, the patrol leader checks the progress of the patrol's preparation. He makes changes in his preliminary plan if necessary. He completes his plan and prepares his patrol order.

    (2) At the time and place designated in his warning order, he meets his patrol and issues the patrol order. After answering questions and satisfying himself that his men understand the mission and their duties, he dismisses them to complete their preparation. He supervises the patrol and conducts any rehearsals that he feels are necessary. When possible, rehearsals are conducted on terrain similar to that over which the patrol is to operate. Before de-

parting, the leader conducts a final inspection. At the designated time, the patrol is moved to the naval docks where the patrol leader meets the commander of the naval vessel involved and makes his final coordination. He has the patrol load rubber rafts and conducts a rehearsal on loading and unloading from the mother vessel. At the completion of the rehearsal, the rubber rafts are loaded and the troops are fed a hot meal. At the designated time, the patrol boards the naval vessel for transportation to the dropoff area. The vessel then moves to a point 1,000 to 2,000 yards off shore from the landing area. The patrol leader has each rubber boat unloaded from the mother vessel and assembled for movement to the enemy shore. Direction of movement is maintained by an azimuth, with the patrol leader in command. A friendly agent from the mainland assists in guiding the incoming boats to the landing area by flash signals. Upon reaching the mainland, the boats are carried ashore and hidden. The agent, in addition to signaling the patrol, gives them a radio. This enables the patrol to call for artillery support at the objective or along the route. The patrol leader moves his patrol over his selected route to the objective, arriving there before daylight. Any one of several objectives may be established for this exercise. The objective should logically warrant this type of waterborne operation. For example, the objective might be a critical heavy artillery position, a guided missile installation, or a radar installation. The mission does not necessarily have to be one involving the destruction of an installation. It is logical to have the patrol release prisoners from a stockade. The evacuation of a wounded general officer by helicopter may be included in the mission. The patrol completes its mission and moves over a selected route to a position designated as a landing zone for helicopters. Here it is resupplied with ammunition and rations, and, if appropriate, the wounded general officer is evacuated. If a mission other than releasing prisoners is assigned, the observer-instructor may declare one of the patrol members a casualty for evacuation.

(3) Aggressor personnel are located near the landing zone but do not make contact with the patrol until ten minutes after the helicopter departs. Contact with the patrol continues until proper action is taken by the patrol leader. After the supplies and rations are distributed, the patrol moves toward a clandestine bivouac area. Here reorganization might take place, if necessary. The men eat their rations and prepare for the final movement to the boats. They then move to a preselected rendezvous point where they once again make contact with the agent. The patrol leader returns the radio to him and has his patrol prepare its boats for the return to the mother vessel. The patrol leader moves his men to the mother vessel after receiving a signal from the ship. The vessel should be located 800 to 1,000 yards off shore. Two

minutes after the patrol receives its signal, fighter aircraft make their first pass at the beach. Demolitions are detonated to simulate bombing. An aircraft drops flares to light up the area. The fighter aircraft strafe the area as the boats move toward the mother vessel. The support of the aircraft, the vessel, and the helicopters, should be coordinated well ahead of time to insure continued realism. When the patrol reaches the mother vessel, it loads the rubber boats and returns to the docks, where trucks are waiting. The men load their equipment and return to the reserve area. Upon arrival, they are debriefed, fed a hot meal, and critiqued by the observer-instructor.

37. General and Special Situation

A general and special situation constituting an operation order is presented by the principal instructor who introduces himself as the S3. The following is a suggested method of presenting the situation for this problem.

a. General Situation.

(1) Ten days ago the (unit) Aggressor Army made an air and waterborne landing at (location). Since that time, they have taken advantage of every situation, and at the present time have a beachhead of about ____ square miles. In our immediate sector their lines run from the coast to here as shown on this overlay. According to the latest intelligence reports, the following units are opposing our division: The ____ rifle division, and the ____ rifle

division. Their location is as shown on this overlay.

(2) Our forces are employed as shown on the overlay.

b. *Special Situation.*

(1) You are at present located as shown on this overlay.

(2) All divisional units have been instructed not to send out any patrols for a period of 36 hours, effective 1800 hours today. They were given no reason. They will not know that you are out to their front.

(3) Your mission is to (state desired mission).

(4) (Rank) (Name) will give you the naval phase of the operation later.

(5) (Include here a statement relating the importance of the mission to the overall operations in the exercise.)

(6) You will make your landing at (coordinates). Here you will meet a friendly agent who will give you information of the enemy. He will guide you onto the beach by means of flashlight signals. Each signal will take place every ____ minutes and will consist of ____ flashes.

(7) You will move to the objective and accomplish your mission.

(8) If the mission includes the release of prisoners and evacuation of the general officer, include here pertinent details to include the helicopter pickup.

(9) After the mission has been completed, move to a clandestine bivouac area and remain there all day tomorrow.

(10) Accomplish the mission before the first light tomorrow morning (time) hours.

(11) At the helicopter landing zone located at (coordinates), serious casualties may be evacuated.

(12) Return to the area of the beach where you landed and again contact the agent.

(13) The agent's name is ____. Challenge him with the code words ____ ____, on making contact. He will reply with the words ____ ____.

(14) You will receive a signal from a naval vessel at ____ hours. This will consist of ____ flashes of a (color) light every ____ minutes. Upon receipt of this signal, launch your boats and move to the vessel. Upon arrival, load your boats and prepare for returning.

(15) Give me your tentative route before your departure from this area. Also include the location of your planned clandestine bivouac area.

(16) Heavy artillery, in addition to navy fire support, can support you at the objective area. Make arrangements for this support with the artillery liaison officer during your coordination.

(17) You will have to adjust your supporting artillery fire.

(18) For a mission of this type, it is necessary to coordinate closely with the Navy. (Rank) (Name) (USN) is here to tell you about the naval phase of this operation.

Naval Phase

I have been detailed by the Navy to work with the Army on this operation. I shall be with you until it is completed. I will be in command of your boat and put you ashore. At 1600 hours today, a power boat will dock at the landing near _____. At that time, the operation supply officer will take you down and point out your boat to you. You may start inflating your boats and stowing your equipment. At 1900 hours, you will have your men and equipment aboard, and the vessel will depart and proceed (give location), at ___ knots for about ___ miles. We should arrive at our rendezvous here (pointing), a mile off the mainland, about ___ hours. There you will debark and the Navy will leave you.

The weather tonight will be _____. The sea will be _____. Darkness is at ____ hours on (date). Daylight is at (time) hours.

Let's talk about your return trip. At 0300 hours on (date) I will be at the same location off the mainland. On your signal, I will come in as close as possible and pick you up. Contact will be by means of a hooded flashlight. Give us a series of three dots. I shall risk an answering signal in the same manner. (The battle group S2 or S3 continues the order at this point.)

(19) Decide what ammunition you want for the mission. There is to be no resupply.

(20) We have been able to obtain a helicopter for the evacuation of your wounded. It will be on a standby basis. The only place it can land is (location).

(21) (Name) (rank) will assist you in planning for the helicopter operation.

(22) Communication will be maintained by AN/PRC-10. One net will be maintained between you, the commander of the naval vessel, and this headquarters.

(23) Everything you need for your mission will be taken from here. The evacuation helicopter will also resupply you with rations.

(24) Call signs are as follows:
- _____ Headquarters
- _____ Patrol
- _____ Naval vessel

(25) Use the same net for requesting supporting fires.

(26) You are reminded that you are in a marshaling area near a POE. You have just been given a very important job. We want the job done and we want to see you come back. So, in an effort to seal any leak of information out of here, guards are posted on the gates. You will not be permitted out of this area without written permission from me. Under no circumstances are you to talk to a civilian today. The naval officer and I will be here all day to answer any questions you may have.

(27) Weather will be _____. Darkness is at ____; daylight, ____.

(28) Are there any questions?

(29) Time is now ____.

c. *Critique Checklist.*

(1) Comment on boat landing plan and selection of clandestine bivouac area.

(2) Comment on debarkation and movement to beach area.

(3) Discuss encounter with friendly agent.

(4) Were all assault boats properly camouflaged?

(5) Were footprints erased from landing point on beach?

(6) Were possible danger areas avoided?

- (7) Did patrol use stealth to gain surprise?
- (8) Were security guards eliminated silently?
- (9) Discuss action at objective, conduct of assault, reorganization, withdrawal.
- (10) Was aerial resupply and evacuation of wounded conducted in a sound, tactical manner?
- (11) Discuss clandestine bivouac area, location, camouflage, security.
- (12) Did patrol slacken its security as fatigue increased?
- (13) Discuss survival techniques.
- (14) Discuss Aggressor action other than at objective.
- (15) Discuss actions of patrol during strafing attack by fighters.
- (16) Discuss withdrawal from hostile shore.

SUGGESTED TIME SCHEDULE

0600–0700	Briefing.
0700–1200	Planning—issue warning order—coordination.
1200–1300	Dinner.
1300–1400	Continue planning.
1400–1500	Issue order—rehearsal—inspection.
1500–1600	Move to embarkation point.
1600–1700	Rehearsal—inflating boats—loading mother vessel.
1700–1800	Supper.
1800–1900	Final preparation—inspection.
1900–2000	Movement by naval craft to unloading area.
2000	Debark in boats.

2d Day

0600–0700	Action at objective.
0700–0800	Movement to bivouac—helicopter operations—movement to landing beaches.
1800–1000	(3d Day) Movement to navy vessel—return to marshaling area—debriefing—critique.

SKETCH OF PROBLEM LAYOUT

Clandestine Bivouac
⋀
1 mile
⋁

Helicopter LZ
⋀
1–2 miles
⋁

Objective
⋀
5–8 miles
⋁

Mainland
⋀
800–1,000 yds
⋁

Debarkation Point
⋀
As required
⋁

Embarkation Point
⋀
As desired
⋁

Marshaling Area
(Reserve Area)

Note. It is recommended that the distance for movement for a patrol in this exercise be approximately ten miles. However, the distance can be scaled to suit the available terrain. A sufficient number of situations should be included to keep the patrol occupied for a period of twenty hours after landing on the beach.

Section XI. RAID AGAINST INSTALLATION DEEP IN ENEMY TERRITORY

38. Purpose

To develop the ability of leaders to prepare a thorough patrol plan and to execute the plan under realistic combat conditions. To familiarize the members of the patrol with the principles and techniques of day and night patrolling. To develop the ability to infiltrate the enemy's battle position and move deep into his rear area to a designated objective. To develop the ability to raid an objective and move to a designated helicopter pickup point.

39. Scope

a. Day and night combat patrol. This is a day and night combat patrolling exercise to include planning, preparation, execution, and critique phases. The problem requires infiltration of the enemy battle position by four groups, movement to a regrouping point five miles to the rear, movement to a rendezvous with a civilian agent, movement to a deep rendezvous point with another civilian agent, movement by boat across a lake, raid of an installation fifty miles in enemy territory, disabling of the installation, and movement to a predesignated helicopter pickup point for return to friendly lines. See paragraph 3.

b. See FM 7–10 and FM 21–75.

40. Scenario

a. This problem is scheduled as the last exercise because it repeats most of the principles taught previously. It contains many situations and requires numerous decisions by the patrol leaders. It thoroughly tests the patrol leader, his chain of command, each individual of the patrol, and the patrol as a team. The size of the patrol on this problem should not be more than 45 nor less than 18 men.

b. The patrol is oriented and the patrol leader is presented with the operation order, to include his patrol's mission. The patrol leader now becomes responsible for the execution of the proper steps in troop leading procedure required for the successful accomplishment of his mission. Generally, the actions of the patrol leader follow the sequence presented below.

    (1) The patrol leader makes an initial estimate of the situation and decides on an efficient way to utilize the time available to him. He makes a thorough map study and forms a preliminary plan. He coordinates with personnel available in the area. There will be present a friendly partisan who has lived close to the objective prior to Aggressor's occupation of it. The commander of the helicopter group who will effect the withdrawal should be present to coordinate with the patrol leader. When his coordination is completed at battle group, the patrol leader returns to his patrol and issues a warning order so that preparation can be started by all members. Then, with selected subordinates, he departs for the forward area through which his patrol is to pass. A special situation imposed upon the patrol leader requires that he divide his unit into four groups in order to infiltrate the Aggressor battle position. The special situation requires that each of the four groups leave from separate forward units, thus requiring co-

ordination with more than one forward unit commander. Upon arrival at the forward lines, he meets the S3 of the battle group, who assists him in coordinating with the forward units through which the patrol is to pass. He and his subordinate leaders coordinate with the company commanders involved and obtain any information that may be available. He and his subordinate leaders also make a visual reconnaissance of each forward unit with which they are each concerned and the initial routes they are to use. The patrol leader and his subordinates then return to the patrol. The patrol leader checks the progress of the patrol's preparation and makes any necessary changes to his preliminary plan. He then completes his plan and prepares his patrol order. At the time and place designated in his warning order, he meets his patrol and issues the patrol order. After answering questions and satisfying himself that his men understand the mission and their duties, he dismisses them to complete their preparations. He supervises the patrol and conducts any rehearsals that he feels are necessary, such as the actions at the objective and method of loading the helicopters. He conducts the rehearsals on terrain similar to that over which the actions will actually take place. Before departing, the leader conducts a final inspection. At the designated time (darkness), the patrol moves forward to a detrucking point behind friendly forward lines. As the patrol detrucks, an artillery barrage falls on the convoy. This is done by means of concealed demolition pits located near the detrucking point and control personnel using simulator artillery ground bursts. Control personnel also lie in and around the detrucking point moaning as though wounded. Each patrol group is met by a guide who takes it to a respective outpost. From this point until such time as the entire patrol rallies at the first march objective, each patrol group is in effect a separate patrol with its own leader. The different patrols then move to the Aggressor lines and infiltrate. Each group leader and the patrol leader move their patrols out and guide them on the prescribed route. Just outside of friendly lines, the leader halts his patrol, establishes security, and listens for signs of any enemy ambush. Satisfied that there is none, the patrol continues. A special situation is imposed on the patrol leader wherein he is required to move his patrol groups through known gaps in the enemy battle position with the ultimate purpose of regrouping at the first march objective, march objective "A." The leaders use pace men and compass men as desired. They change the formation as the terrain or proximity to known enemy positions dictates.

(2) When the groups approach the enemy lines, the group leader halts the patrol and sends a reconnaissance team forward to reconnoiter the area and locate the gap. If undetected, the patrol moves through the gaps, utilizing maximum stealth. Once behind the Ag-

gressor main battle position, the patrol continues to march objective "A." The leader halts the patrol and has it take cover in place if he feels that the enemy hears him, but he does not yet know his exact location. If satisfied that the enemy is yet unaware, the patrol leader may have his group continue. If he feels the enemy is alerted to his presence, he may withdraw and attempt an infiltration at another location. If detected within the main battle position, the patrol leader may have his patrol fight its way through until contact is broken, at which time he continues on with his mission.

(3) The patrol groups should rally at march objective "A" just after daylight the following morning. The first group to arrive establishes security and awaits the arrival of the other groups. A partisan arrives at this position and the patrol leader or senior group leader present coordinates with him concerning information of the enemy and the location of food for the patrol. During the day, the patrol establishes security, cleans weapons and equipment, and rests. The patrol leader issues any subsequent orders that he may find necessary. All of his men are accounted for, all wounded are treated, and generally, he reorganizes after the previous night's encounter with the enemy. It is desirable that "A" be located next to a natural obstacle such as a large river. This makes the entrance of the partisan into the exercise more logical in that he can give the patrol informa-

tion concerning a crossing site. Bridges crossing the obstacle would therefore be guarded by Aggressors. The presence of the obstacle and the special situation of the partisan having the only knowledge of a crossing site or fording point, will enable the principal instructor to prevent the patrol's movement from the area prior to darkness. This facilitates control of the Aggressor action to follow, since the location of the patrol is known throughout most of the day. Just after darkness, the partisan guides the patrol to a hidden cache of food. The patrol leader distributes the food and then the patrol moves out. Again this can be accomplished either by use of four groups or as an entire patrol, depending on the patrol leader's decision. The next objective, march objective "B," is a control point for the second night's movement. There is no action at "B." It may logically be placed in the operation order as a general guiding point to prevent friendly aircraft from strafing or bombing the patrol accidentally.

(4) The patrol arrives at march objective "C" early in the morning of the following day. Here again it contacts a friendly partisan who supplies it with food. It is here that final plans are made for the accomplishment of the mission. Reorganization accomplished, the patrol rests. If possible, this point should be located near a large body of water, thus necessitating the patrol's use of small boats in a successful crossing. The boats could be

supplied by the partisan. If boats are used, the leader is told in the operation order that they are available. This will necessitate his having a boat loading plan. It also presents various problems to be solved by the patrol leader, such as control and security during movement across the body of water, and methods of embarking and debarking.

(5) At darkness the patrol moves out as a group toward its objective. It is desirable that the boats, if used, be employed in the initial movement from "C." On arrival at the attack position, the patrol leader halts the patrol, and with selected subordinates he goes forward for a reconnaissance of the objective. After completing his reconnaissance, he and his selected subordinates return to the patrol. He makes a final check of the patrol and issues any further orders he feels necessary to accomplish the mission. He moves the patrol toward the objective, establishes his security and support teams, and sends his killer team into the installation. He supervises the placing of demolitions to insure the mission is a success. He supervises the withdrawal of his attacking force when he is satisfied the mission is accomplished. He controls the assault force and regroups the entire patrol at a predesignated assembly area. He then moves the patrol to march objective "E," the helicopter pickup point. He may have his patrol move to "E" in small groups or teams rather than as one unit.

(6) The Aggressor action at the installation should be normal for the situation. Walking guards may be placed around the installation, reserves may be withheld until the patrol launches its attack. If the patrol has not compromised its mission to the enemy prior to the attack, then the Aggressor troops should react normally—that is, they should not be lying in wait for the patrol. If, however, the patrol has compromised its mission somewhere along the route (for instance, a member may have been captured with a marked map showing the objective), the Aggressor force should ambush the patrol at the installation. It is desirable to allow the patrol to be ambushed, since it affords realistic training. After the attack, the enemy should pursue the patrol, taking advantage of every opportunity to ambush and capture to prevent its movement to the helicopter pickup point and final withdrawal. The patrol should arrive at point "E" early the following morning. Security is established and final preparations made for the pickup. If helicopters are available for the exercise, the pickup is accomplished as planned. If the helicopters are not available, the observer-instructor should have the patrol simulate loading. Aggressor action should take place at "E" in the form of an attack on the patrol. It is most desirable that this action occur at the moment of arrival of the first helicopter or, when no helicopters are available, on signal from the observer-instructor that the patrol is simulating loading.

(7) After returning to the reserve area, the patrol leader is debriefed. In order to provide training for all personnel, the patrol leader may be debriefed in the presence of the entire patrol. The patrol leader makes his report and the observer-instructor conducts his critique.

41. General and Special Situation

A general and special situation constituting an operation order is presented by the principal instructor, who introduces himself as the battle group S2 or S3. The following is a suggested method of presenting the situation for this problem.

a. General Situation. The enemy in our sector has been pushed back to the line indicated on this situation map. However, he was able to hold the western sector. (At this point insert intelligence related to the objective installation—for example: "His capability of launching guided missiles assisted him in holding the western sector." This would relate to an installation such as a guided missile site which would, later in the order, be assigned as the objective for the exercise.)

b. Special Situation.
    (1) Reports from the forward companies indicate that the enemy has strong positions; however, our patrols have located several gaps in their line through which small groups can infiltrate by using extreme stealth.
    (2) (Insert here information concerning the natural obstacle located at march objective "A." Also include information that the enemy is using the surrounding terrain as a training

area. This information is given so that Aggressor operations in the area can be realistically injected into the play of the exercise during the second day.)

(3) Enemy alert platoons are located in (logical areas near the objective) on each side of and to the rear of the (objective site). Motorized patrols move between these locations every (give time schedule) hours. A few enemy soldiers are on walking guard around the (objective). There is about a squad located within the (objective) itself. They are probably reliefs for the walking sentries.

(4) Our lines run ____ as you can see on this overlay. We have no operations planned for the next three days.

(5) Your mission is to raid the (installation) on the night of (date) prior to 2400 hours.

(6) Your objective is located ____ miles behind the enemy lines.

(7) In order to infiltrate the enemy lines, divide your patrol into four infiltration teams. Each team will leave from a different forward outpost. The patrol leader will go to ____ company at 1400 hours this afternoon to coordinate movement through the forward area. The S3 of the forward battle group will be there to assist you. He will give you the location of the gaps in the enemy lines.

(8) Cross the line of departure at (darkness) hours this evening.

(9) You will follow, generally, the route indicated on the overlay provided for you. The overlay shows your march objectives "A," "B,"

"C," the objective, and march objective "E." After accomplishing your mission, proceed to the vicinity of march objective "E" but do not physically occupy it.

(10) At ____ hours on (date) you will occupy point "E" and prepare for a helicopter pickup. There will be (number) helicopters and each will carry ____ members of your patrol. If helicopters do not arrive at the designated time, wait fifteen minutes, then return to a bivouac area and wait (time). You will again be at point "E" for pickup. If the helicopters do not appear in fifteen minutes, make your way back to friendly lines as best you can. Captain ____ of the helicopter company will be available for coordination and information during your preparation.

(11) A civilian named ____ ____ will try to contact your patrol at march objective "A" at ____ hours tomorrow. He will identify himself with the code word "____." He will have information of the enemy activity and a resupply of food.

(12) (If a natural obstacle is available at point "A," mention pertinent information concerning it; for example, width, depth, rate of current, if it is a stream. Also mention that the civilian (name), since he has lived in that area, will probably give information concerning the best location for crossing the obstacle.)

(13) Guide generally on march objective "B." This will prevent the patrol from getting into areas subjected to bombing and strafing from friendly aircraft.

(14) A civilian named ____ ____ will try to contact you at march objective "C" at 0700 hours on (date). He will wait for your arrival and will have information concerning the enemy and a resupply of food. He will also identify himself by the code words "____ ____." (If there is a large body of water at point "C," necessitating the use of small boats in crossing it, this civilian can also supply boats. If this is the case, give that information at this point in the order.)

(15) Start for your objective from "C" after dark on (date). Accomplish the mission prior to 2400 hours.

(16) Request and draw your equipment and ammunition requirements through your company supply.

(17) No rations will be carried.

(18) Transportation will be furnished by the battle group. Depart from your reserve location tonight at ____ hours.

(19) Patrol leader will make necessary disposition of casualties and prisoners. There will be available during daylight hours, a helicopter for medical evacuation. You can request it as needed.

(20) You will take an AN/PRC-10 for communication with this headquarters. Use it only after accomplishing your mission or after the element of surprise is lost. Primary channel ____. Alternate channel ____. Your call sign is ____. This headquarters' call sign is ____.

The challenge and password for each day you will be out is—

1st day _____
2d day _____
3d day _____
4th day _____

(21) Weather will be _____. Darkness is at _____, daylight, _____.
(22) Are there any questions?
(23) The time is now _____.

42. Personnel and Equipment Requirements

a. Personnel. In addition to the assigned principal instructor, assistant instructor, and the appropriate administrative personnel, the following are the minimum requirements for each lane on this problem.

(1) One officer per patrol (observer-instructor). It is desirable to have one observer-instructor with each of the four patrol groups until they regroup at point "A."

(2) Three EM on each friendly forward position—company commander, guide, and outpost man.

(3) One officer at the forward position—battle group S3 on line.

(4) Twenty EM as enemy at battle position; five located near each one of the four gaps.

(5) Two NCO's act as friendly partisans after the first night of the problem. The enemy representation at the battle position can be moved to point "A," and the following day, to "C." During the night of the attack, all Aggressors could be utilized at the objective reserve

areas. The final morning of the exercise, they can all be employed in the attack at "A."

b. Equipment. Normal individual equipment should be provided. Aggressor uniforms and materials for preparation of positions are issued as available. Radio, telephone, and demolitions may be used as described. Pneumatic boats or similar type craft are desirable. If they are used, life preservers, either inflatable or kapok, should be provided. Helicopters, if available, should be provided.

c. Critique Checklist.

    (1) Were additional security measures taken as patrol neared Aggressor MLR?

    (2) Did patrol infiltrate MLR in a tactically sound manner?

    (3) Were danger areas avoided when possible?

    (4) Were additional precautions taken as patrol neared march objective "A"?

    (5) Discuss actions at clandestine bivouac to include coordination with and interrogation of friendly partisan, perimeter defense, cleaning of weapons and equipment, security and rest.

    (6) Discuss movement from "A" to "C."

    (7) Discuss actions at second clandestine bivouac area to include reconnaissance of objective area, rehearsal and final planning, and orders.

    (8) Discuss actions at objective, to include use of stealth and tactical plan.

    (9) Did patrol become careless of its security as fatigue increased?

    (10) Discuss movement to "E."

(11) Discuss actions at "E," to include loading plan, defense plan, time of occupation, marking of LZ, signals to helicopter, action during enemy contact.

SUGGESTED TIME SCHEDULE

0730–0900	Briefing.
0900–1200	Planning by patrol leader.
1200–1300	Dinner.
1300–1430	Planning continued.
1430–1515	Warning order issued by patrol leader—prepare for reconnaissance.
1515–1630	Reconnaissance.
1630–1915	Patrol order issued—supervise final preparations—supper—rehearsal—inspection.
1915–2030	Movement to LD.
2030	Cross LD.

SKETCH OF PROBLEM LAYOUT

March Objective "E"	———————	———
	↕ 7–8 miles	↑
Objective Installation	———————	3d night
	↕ 2–3 miles	
Natural Obstacle (if possible)	———————	
	↕ 200–300 yds	
March Objective "C"	———————	↓
	↕ 4–5 miles	↑
March Objective "B"	———————	2d night
	↕ 4–5 miles	
Natural Obstacle (if possible)	———————	
	↕ 200–300 yds	
March Objective "A"	———————	↓
	↕ 8–10 miles	↑
Enemy Battle Position	———————	
	↕ 1,000–1,500 yds	1st night
Outpost (L.D.)	———————	
	↕ 50–200 yds	
Forward Command Post	———————	
	↕ As desired	
Reserve Area	———————	↓

Note. It is recommended that the distance for movement for a patrol on this problem be approximately fifty miles. However, the distance can be scaled to suit the available terrain. A sufficient number of situations should be included to keep the patrol occupied for a period of 72 hours after departing the outpost.

APPENDIX V
PATROL TIPS

1. **Planning and Preparation**

   a. *Patrol Order Tips.*

   (1) Assign every member of patrol an area of responsibility. Do not forget to appoint a man to observe overhead and to the rear.

   (2) Have more than one man designated as a pacer, and use the average pace obtained from both.

   (3) Use the point man as a point and not as a compass man; he is primarily concerned with security. Have the second or third man responsible for navigation. On large patrols point squads may be used.

   (4) On small patrols the second in command should send the count forward after each extended halt or passage of an obstacle. On large patrols use the chain of command to account for men.

   (5) In giving the patrol order, visual aids are of great value. Use a blanket board, blackboard, sandtable, or even just a sketch using a stick and a cleared piece of ground.

   (6) Make a good map reconnaissance; know your route from memory prior to departure.

   (7) Consider the use of seemingly impassable terrain in planning your route. You are less likely to encounter the enemy. Insurmountable natural obstacles are very rare.

   (8) Avoid all human habitations.

(9) In mountainous terrain plan to utilize ridge lines for movement whenever possible. On bald mountain tops or ridge lines with sparse vegetation, avoid being skylined.

(10) In planning a route, do not forget to use offsets when applicable. An "offset" is a planned magnetic deviation to the right or left of the straight-line azimuth to an objective. It is used to verify your exact location (either to the right or left) in relation to the objective (fig. 14). An accurate means of planning an "offset" is to lay off the desired offset distance on the map using the graphic scale, and to determine the grid azimuth to the desired "offset" point, finally, converting the grid azimuth to magnetic azimuth.

(11) In issuing the patrol order give specific action to be taken at each danger area as determined by the patrol leader.

(12) When enemy wire is encountered, cut only when necessary. Make a proper reconnaissance first.

(13) There are several acceptable methods of crossing roads. Whatever the method used, the basic principles of reconnaissance and security apply. Some of the accepted methods are—
  (a) Patrol can form a skirmish line and advance across the road on the double.
  (b) Entire patrol can form a file, following the man's footsteps in front to minimize detection of footprints.
  (c) Men cross the road a few at a time until patrol is across.

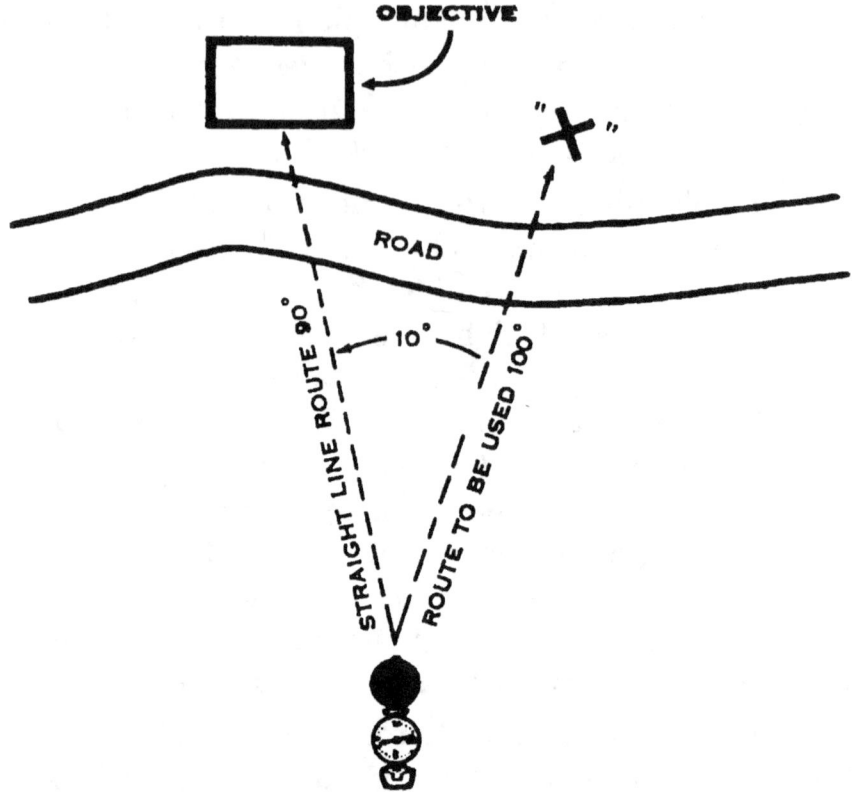

Figure 14. Use of offset. Remember the distance from "X" to the objective varies directly with the distance to be travelled and the number of degrees offset.

(14) Crossing streams is similar to crossing roads; both reconnaissance and security are necessary.

(15) Select check points from map reconnaissance prior to departure, and confirm their location on the ground as rally points as patrol passes over them.

(16) When necessary to infiltrate enemy lines, a rally point should be preselected behind enemy lines. An alternate point should also be selected in event the first point is occupied by the enemy.

(17) When preparing an equipment list, consider fragmentation, white phosphorous, concussion, smoke, luminous, and thermite grenades.

(18) Light automatic weapons are good for night patrols. Avoid taking several different types of weapons on patrol, as it makes ammunition redistribution difficult.

(19) Test fire automatic weapons prior to departing on patrol to insure their proper functioning.

(20) Develop small individual survival kits and carry them on all patrols.

(21) A length of rope, which can be easily carried secured around the waist, has many uses on patrol: securing prisoners, aiding in climbing or descending obstacles, crossing rivers, etc.

(22) Carry gloves to protect hands from briars and scratches.

(23) Blackjacks can be made out of wet or dry sand, soap, or stone-filled socks.

(24) A garrote can be used for killing a sentry or capturing a prisoner. Use an insulated wire if you want to capture a prisoner.

(25) Keep the cutting edge of the entrenching tool extremely sharp. It is a good silent weapon and can be used in lieu of a machete.

(26) A razor blade or sharp knife and a piece of cord is a good substitute for a snake bite kit.

(27) A candle or can of sterno placed under a poncho is a great aid in keeping warm, particularly if used in conjunction with a deep slit trench.

(28) Two pieces of luminous tape, each about the size of a lieutenant's bar, worn on the back of

the collar aid in control and movement on dark nights. Turn collar down if near the enemy. Pieces of luminous tape can also be worn on the cap but reverse the cap if near the enemy.

(29) Both in day or night carry and use binocular whenever practical.

(30) Take one or more ponchos on patrol; they can be used as litters, for constructing rafts, to conceal lights, and as shelters.

(31) Carry extra socks on person at all times.

(32) Carry a sharp knife. It is best carried concealed in the boot or shoulder holster.

(33) Carry two of each of the small items of equipment such as fuse crimpers and wire cutters.

(34) Carry additional batteries for both flashlight and radio, especially on long patrols.

(35) Consider use of scout dogs.

(36) Suspenders should in most cases be worn when wearing the cartridge belt, AR belt, or pistol belt.

(37) Always carry cleaning equipment for individual weapons, regardless of the length, type, or nature of the mission. Check to see that the oiler is full. Patches carried in the stock of the rifle prevent cleaning equipment from rattling.

(38) Tape rifle sling to weapon to prevent noise and snagging. Slings may be used to secure splints and as tourniquets. Insure that the tape on weapon does not hinder the operation of the piece.

(39) Soot, paste, and other types of camouflage material should be used freely. Attention should be given to all exposed skin, including the back of the neck, behind the ears, and the backs of the hands.

(40) A clear acetate sheet placed over luminous tape can be used to make rough strip maps at night. The map will glow. Use a grease pencil because any information can be easily erased.

(41) Light machine gun ammunition, minus the box, can be carried conveniently in the combat pack suspended from the chest. If necessary, it can be fed into the gun from the pack.

(42) If necessary to leave a wounded man to be picked up later, leave another man with him if possible. When in close enemy contact, remove wounded from immediate danger areas before treating.

(43) All signals to be used on patrol should be prearranged and known to all members. Keep signals simple and to a minimum.

(44) Avoid using the password forward of friendly lines. Plan an alternate password or signal to be used there.

(45) Sound signals, such as taps on the rifle butt, are practical when used in small patrols but are impractical when used in large patrols.

(46) Over short distances such as the width of a road, the compass can be used for signaling at night. A piece of luminous tape can also be used for this purpose.

b. Miscellaneous.
  (1) When possible, arrange to have a light aircraft reconnoiter ahead of your patrol to keep you informed of any enemy activity or ambushes along your route.
  (2) When a reconnaissance is to be made, the patrol leader or designated representative should be accompanied by at least one other responsible man.
  (3) Prior to arrival at the command post or outpost through which you plan to pass, have a list of coordinating questions prepared.
  (4) Coordinate fully with outpost personnel through whose position you are departing.
  (5) Hold a rehearsal on terrain similar to that on which you will later be operating. Cover all details.
  (6) Insure that all equipment is checked before departing.
  (7) Fold and prepare maps before leaving to facilitate map checks while en route.
  (8) Preset compasses prior to departure. This minimizes confusion, delay, and inaccuracy—particularly at night.
  (9) Check to see that grenades carried can be reached easily.
  (10) Allow approximately thirty minutes for your eyes to become accustomed to the dark before departing on a night patrol.

2. Execution
  a. Movement.
    (1) When moving at night take advantage of any noises such as wind, vehicles, planes, shelling,

battle sounds, and even sounds caused by insects.

(2) Stay off roads and trails for movement unless their use is deemed absolutely necessary.

(3) Guide on terrain features.

(4) Use stars to aid in navigation. When doing so, confirm your location periodically with a compass.

(5) The night method of using the compass can often be used during daytime to facilitate movement in dense terrain.

(6) When in close proximity to the enemy main battle position, avoid lateral movement across its front and rear.

(7) Consider use of supporting weapons to aid in navigation. Use artillery, mortar, .50 caliber, or recoilless marking rounds.

(8) Use the "flash-bang" method to determine your distance from the impact area when using marking rounds, in order to fix your location in relation to the objective.

(9) When men cannot stay awake on security and at halts, minimize the number of halts and make the men assume a kneeling rather than prone position.

(10) Weapons should normally be carried at a ready position.

b. *Miscellaneous*.

(1) Never throw trash on ground while on patrol. Bury and camouflage it to prevent detection by the enemy.

(2) During halts at night in terrain in which control is difficult, halt the patrol in place, face in the direction of responsibility, and kneel.

(3) Between main battle positions or whenever in close proximity to the enemy, there should be absolutely no smoking. Behind the enemy main battle position smoke only with caution. Control smoking day and night.

(4) Do not jeopardize security by letting ear flaps and hoods interfere with the hearing ability of the patrol.

(5) When on patrol pass simple instructions; give time for dissemination; then execute.

(6) Keep talking to a minimum. Use arm and hand signals.

(7) When reconnoitering enemy positions, keep covering force within supporting distance of reconnaissance element.

(8) Do not mark maps with friendly information. Maps may be marked with enemy information.

(9) When possible while on long patrols, allow men to sleep; but maintain proper security.

(10) Never take the entire patrol to make contact with friendly agents or partisans. Have one man make the contact and cover him.

(11) Know the method of finding the North Star; know the watch and sun method of finding North.

(12) The best nights for patrols are dark, rainy, and windy nights.

(13) Do not let the importance of personal comfort endanger the patrol and the accomplishment of the mission.

APPENDIX VI
INSTRUCTOR'S GUIDE
(RANGER HISTORY)

1. Organization

The history of the American Ranger is a saga of courage, daring, and high leadership. It is a story of men whose skills in the art of fighting have seldom been excelled.

a. Robert Rogers (Major). The first rangers were organized in 1756 by Robert Rogers, a resident of New Hampshire, who recruited a total of nine companies from among the continentals. The British Army at this time was fighting the French and Indians (Seven Years' War). Rogers' men were skilled in woodland warfare and were able to travel great distances over difficult terrain. As the chief scouting arm of the British, they were bold in procuring intelligence by scouting enemy forces and positions and capturing prisoners. Rogers' fighters accompanied Wolfe's expedition against Quebec in the Montreal Campaign of 1760, and participated in the Western campaign as far as Detroit and Shawneetown to receive the surrender of all French outposts. In the west in 1763, Rogers and his men distinguished themselves in the Battle of Bloody Ridge.

b. Daniel Morgan (Colonel).

    (1) The type of fighting evolved by Rogers and his men was developed during the Revolutionary War by Colonel Daniel Morgan, who organized a unit known as Morgan's Riflemen.

These men, clad in frontiersman's buckskin garb, schooled in the Indian's methods of forest fighting, and armed with the deadly accurate frontiersman's rifle, were without equal. According to General Burgoyne, Morgan's men were "... the most famous corps of the Continental Army, all of them crack shots..."

    (2) Morgan's Riflemen fought at the Battle of Freemans Farm (Sep 1777), and at the Battle of Cowpens (Jan 1781), where they inflicted heavy losses on the main body of British troops commanded by Colonel Tarleton. These successes were in large part due to the proper use of natural cover and surprise tactics.

c. *Francis Marion* (Colonel).

    (1) Another famous Revolutionary War element was organized and led by Francis Marion. Marion's Partisans, numbering anywhere from a handful to several hundred, operated both with and independently of other elements of General George Washington's army. By disrupting British communications and preventing the organization of loyalists to the British cause, they, along with the forces of Sumter and Greene, contributed materially to Colonial victory.

    (2) Marion's group took part in the capture of Fort Johnson and in the victory of Charleston (1775), which gave the southern states a respite from fighting for nearly three years. Again active in 1780, Marion failed in an attempt against Georgetown, but his group was instrumental in the capture of Fort Watson

and Fort Motte, South Carolina, the following year. The loss of Fort Motte—on the line of communication between Camden and Charleston—was a great blow to the British cause.

(3) A favorite retreat of Marion's fighters was Snow's Island. Deep swamps bordered the island, and inland great quantities of game and livestock existed. Marion's men were able to launch sudden attacks from the island in any direction, surprising, killing, or capturing bands of Tories gathering to aid the British. After each action they withdrew to the safety of the swamps. A British colonel once pursued Marion's band through these swamps for 25 miles before being forced to a halt by a seemingly impassable swamp. Angered, the colonel cursed, ". . . the damned fox, the devil himself could not catch him. . . ." Marion was known thereafter as the "Swamp Fox."

(4) Marion's men were good riders and expert shots. They kept close watch on the British, and they continually raided outposts, scouting parties, and lines of communications. There was no certain defense against them, and their activity necessitated the presence of British regulars even in conquered regions. This organized partisan activity was most successful against an enemy of superior forces and discipline.

2. Civil War

The War between the States was again the occasion for the creation of specially trained units such as rangers. The Confederacy quickly capitalized on their usefulness by authorizing the formation of partisan units. The Union forces did not employ ranger tactics until the summer of 1863, and then only on a limited scale.

a. John S. Mosby (Colonel).

(1) John S. Mosby, a master of the prompt and skillful use of cavalry, was one of the most outstanding Confederate rangers. He believed that by resorting to aggressive action he could force his enemies to guard a hundred points while he waited to attack any he chose.

(2) The first real success of Mosby's Rangers was at Fairfax Court House in Virginia, located well behind Federal lines. Mosby had learned that the enemy kept cavalry, infantry, and artillery there. He also knew the officer in charge was Colonel Percy Wyndham, a British soldier of fortune fighting for the Union cause. Mosby's plan was bold—to infiltrate the Federal lines and pluck the officer from the midst of the thousands of soldiers protecting the roads west of Washington. His hope for success was based on the theory that to all appearances it was an impossibility.

(3) Under cover of darkness, Mosby and twenty-nine of his raiders infiltrated Federal outpost lines in the woods north of Centerville. They cut the telegraph lines between Fairfax and Centerville to prevent warning signals from

being sent. The small band reached the outskirts of Fairfax early in the morning. As Mosby had hoped, the Federal headquarters, confident of safety so far behind the lines, was lightly guarded. Mosby and part of his command proceeded to a dwelling which was thought to be Wyndham's headquarters. Meantime, Mosby learned that a Federal soldier whom they had captured earlier was a guard at the headquarters of Brigadier General Edwin H. Stoughton. Directing part of his detachment to Wyndham's quarters, Mosby himself took several men and set out to take General Stoughton. Posing as Federal couriers, they gained entrance into the general's quarters and captured the sleeping officer. The detachment detailed to capture Wyndham reported that the Colonel had gone to Washington the afternoon before; however, they had raided his quarters and had captured the assistant adjutant general, a captain.

(4) It was an unparalleled exploit. Twenty-nine men under a bold and aggressive leader had infiltrated through strong enemy lines to the very point where enemy officers slept, had yanked them out of bed, laughed at their guards, and disappeared before morning. They had captured a general, members of his staff, and many other prisoners and a large number of horses. For over an hour the rangers remained in the town and not a shot was fired!

(5) In March 1863, Mosby defeated a much larger force of Federal troops near Chantilly, Va. An attack which he had planned miscarried, but Mosby's Rangers moved into a half-mile stretch of woods when pursued by Federal cavalrymen. From concealed positions, they delivered deadly carbine and pistol fire into the front and flanks of their pursuers, killing five and wounding several others. One officer and thirty-five men, as well as a large number of horses, were captured. None of Mosby's men were wounded.

(6) At the Miskel farm in the northern tip of Loudoun County, Va., Mosby and his band of sixty-nine men were surprised by a force twice their size. During the bloody fight that ensued in the farmyard, Mosby rallied his men by shouting encouragement above the noise of the turmoil. His men heard and delivered the stroke that brought them victory. The results as Mosby stated to Jeb Stuart were "... nine of them killed—among them a captain and a lieutenant—and about fifteen too badly wounded for removal; in this lot two lieutenants. We brought off eighty-two prisoners..."

(7) Mosby's men were mustered into the regular Confederate service for the remainder of the war. Initially, they formed Company A, 43d Battalion, Partisan Rangers. This unit was a part of the 1st Virginia Cavalry.

(8) During the remainder of 1863, the rangers were busy destroying wagon trains, capturing supplies, horses and mules, and obtaining in-

formation of Federal troop movements and dispositions. From May to July 1864, Mosby's men continued to plague Federal supply and ambulance trains, capturing many prisoners and large quantities of equipment. From one of the most successful of these raids, Mosby emerged with 200 prisoners, 500 to 600 horses, nearly 200 beef cattle, many valuable stores, and $112,000 in payrolls.

(9) In the fall of 1864, an attempt by Federal troops to build a railroad from Manasses Gap westward had to be abandoned because of Mosby's crippling raids.

(10) Mosby was able to preserve and build up his organization over a two and a half years' period within a few miles of the enemy capitol. Numbered among his forces were men who knew practically every road and trail in Virginia and the location of the homes of many Confederate sympathizers behind the Federal lines. They struck in daylight, in darkness, whenever and wherever they could employ the element of surprise. Usually his forays were accomplished with from 12 to 80 men because these small groups could more easily remain concealed and move about unnoticed.

b. *John Hunt Morgan* (General).

(1) Another prominent ranger type unit was the cavalry squadron organized and led by General John Hunt Morgan. Morgan and his Confederate raiders began their famous attacks in December 1861. Their initial attack was on

Lebanon, Ky., 60 miles from Morgan's camp. During this raid they destroyed large quantities of stores and captured several prisoners. A railroad bridge of military importance was burned, thus delaying the movement of Federal supplies to the front.

(2) One of Morgan's most successful raids began in the summer of 1862. With his command of about 800 men, he left Knoxville, Tenn., and made his way westward to Sparta, Tenn., encountering only a few scattered enemy along the way. Turning north at Sparta, Morgan crossed into Kentucky and captured a small garrison, taking 400 prisoners and valuable stores, including enough rifles to equip most of his unarmed men. The raiders then moved on to Glasgow and captured the garrison there, destroying more public stores. These two encounters were typical of the other raids Morgan conducted throughout his two and a half weeks' march behind Union lines. During this time he swelled his own ranks to 1,200 by recruiting en route, marched more than a thousand miles, captured seventeen towns, destroyed millions of dollars worth of Federal stores, dispersed many of the Home Guard, and paroled nearly 1,200 regular Union troops. The losses to Morgan's force in both killed and wounded were less than ninety!

(3) The most famous of Morgan's raids started in July of 1863. With a command of 2,400 men he attacked at Green River Bridge, Ky., but after a severe fight he was forced to withdraw.

Proceeding to Lebanon, Ky., Morgan's rangers captured that garrison. Continuing to the Ohio River near Brandenburg, they crossed on two captured steamers after dispersing hostile troops on the far side. They encountered some militia at Corydon, Ind., but quickly scattered them and captured the town. By this time the entire countryside had risen in arms against them. Newspapers proclaimed an "invasion of Indiana." Reinforcements were hurried in and gunboats on the river were rushed to intercept the Confederate marauders. Following a course roughly parallel to the Ohio River, bypassing Cincinnati, Morgan's men came to within a day's ride of Lake Erie—the deepest penetration of any Confederate force during the war. Near the end of July in the vicinity of East Liverpool, Ohio, Morgan was forced to surrender.

(4) In spite of Morgan's surrender, the raid was successful. It drew off some of the forces which might have harassed Bragg's retreat from Middle Tennessee and which might have helped Rosecrans at the battle of Chickamauga later on.

3. World War II

From the conclusion of the Civil War until World War II there were no ranger units or ranger type training conducted in the United States Army. During both the Spanish-American War and World War I, no ranger organizations came into being. With America's entry into World War II, however, ranger units came forth once again to add to the pages of history.

a. William O. Darby (Major, later Brigadier General). Darby organized the 1st Ranger Battalion during the early part of 1942 in North Ireland. The members of this battalion were all handpicked volunteers. Six officers and 44 enlisted men of the battalion accompanied Commando troops in the Dieppe raid on the northern coast of France. These men learned much of the German's fighting methods and defenses which proved of inestimable value to the Rangers in later operations. The 1st Ranger Battalion participated in the initial North Africa landing at Arzew, Algeria, and the Tunisian battles, where they executed a number of hazardous night attacks over difficult and treacherous terrain. The battalion was awarded the Presidential Citation for distinguished action, which included operations in the critical battle of El Guettar.

 (1) The 3d and 4th Ranger Battalions were organized and trained by Darby in Africa near the close of the Tunisian Campaign. These three battalions made up what was known as the Ranger Force.

 (2) Darby's Ranger Battalions spearheaded the Seventh Army landings at Gela and Licata during the Sicilian invasion and played a leading role in the subsequent campaign which culminated in the capture of Messina. In the Salerno engagement on the Italian peninsula, the Rangers fought for 18 days to hold Chiunzi Pass against eight German counterattacks; in the Venafro battles, they experienced fierce winter mountain fighting in clearing the entrance to the narrow corridor

leading to Cassino. At Anzio the Rangers had the mission of overcoming beach defenses, clearing the town, and forming a defensive perimeter.

b. *Special Service Force.*
   (1) After remaining on the beachhead for 67 days, the 4th Battalion and the remnants of the 1st and 3d were ordered to return to the United States. Only those rangers who had joined originally, and at Arzew, Algeria, were returned. All others were transferred to the American-Canadian Special Service Force which was engaged in holding the lower stretches of the Mussolini Canal facing Littoria and which shortly thereafter joined in the march on Rome.
   (2) The Special Service Force, like the Rangers, was a highly trained volunteer unit that specialized in night raiding and beach landings. It had led the American drive into Kiska in the Aleutians. In Operation ANVIL it spearheaded the landings in Southern France; and later it fought with the Seventh Army near Belfort Gap.

c. *Battle History* (Europe).
   (1) The 2d and 5th Ranger Battalions participated in the D-day landings (6 June 1944) on Omaha Beach. Attached to the 116th Infantry, 29th Division, companies D, E, and F of the 2d Ranger Battalion accomplished their mission of capturing the Pointe du Hoe German coastal battery. The two battalions then assisted in the capture of Grandcamp

and the mopup of scattered enemy opposition between Grandcamp and Isigny.

(2) The 5th Ranger Battalion participated in operations in the Bay of Brest area. Operating on the left flank, they assaulted and captured three of the numerous defenses extending seven miles to Recouvrance.

(3) Later in September 1944, the 2d Battalion, attached to Task Force Sugar of the 29th Infantry Division, drove through numerous outpost strong points to reach the German main line of resistance. The Le Conquet Peninsula was the next objective of Task Force Sugar. The 2d Battalion assisted in this by breaking into the 280-mm gun positions (Batterie Graf Spee) and forcing the surrender of the Le Conquet Garrison commander and 814 prisoners. The 5th Battalion met little opposition in the reduction of the Le Conquet Peninsula defense.

(4) During the Rhineland campaign, 6 to 8 December 1944, the 2d Ranger Battalion, operating in the Hurtgen Forest, successfully captured critical heights near Bergstein, creating a salient in the German lines. Although counterattacked five times and subjected to continuous artillery fire, the unit held the ground which offered observation of not only the key town of Schmidt, but also the Roer River dams. The salient created by the attack reached the most easterly point to which the Allies had driven.

(5) In November 1944, General Patton assigned the 5th Ranger Battalion to XX Corps. A

force consisting of the 6th Cavalry Group and the Ranger Battalion had the mission of screening the XX Corps south flank.

(6) The 5th Ranger Battalion, attached to the 94th Infantry Division, in February to March 1945, accomplished a mission of great consequence to the success of the Allied operations in the Saar River area. Seizure of its assigned objective aided the armored breakthrough which overran Trier and brought elements of the XX Corps to the banks of the Rhine River.

d. Battle History (Philippines). The 6th Ranger Battalion operating in the Pacific formed the nucleus of a rescue force which liberated American and Allied prisoners of war from the Japanese stockade at Pangatian, the Philippines, in January 1945. They moved 29 miles into enemy territory, obtained full support of local civilians and guerillas, and determined accurately the enemy's dispositions. They crawled nearly a mile through flat and open terrain to assault positions, destroyed a Japanese garrison nearly double the size of the attacking force, and, in the dark, assembled over 500 prisoners of war. The prisoners were evacuated from the stockade area within twenty minutes after the assault began. In this action more than 200 enemy troops were killed; Ranger losses were 2 killed and 10 wounded.

4. Korea

a. With the outbreak of hostilities in Korea in June 1950, the need arose once again for rangers. On 25 August 1950, at Camp Drake, Japan, the 8213th

Army Unit was organized from volunteers in the Far East. The 8213th was referred to more informally as the 8th Army Ranger Company and was attached to the 25th Infantry Division. The company participated in the "drive to the Yalu" and spearheaded "Task Force Dolvin" and "Operation Killer." The company was deactivated in March 1951.

b. Fourteen Airborne Ranger Companies were formed and trained at the Ranger Training Command, Fort Benning, Ga., between September 1950 and September 1951.

c. The 1st, 2d, 3d, 4th, 5th, and 8th Ranger Infantry Companies (Airborne) were assigned to divisions throughout the Eighth Army in Korea. These units were deactivated in September 1951 so that ranger trained personnel could be spread throughout the entire infantry in an effort to raise the spirit and overall efficiency of the infantry.

5. Ranger Department

In September 1951, the Ranger Department of the United States Army Infantry School was established at Fort Benning, Ga., with the mission of designing a course to train small unit infantry leaders in leadership and command problems in the field, to improve their physical condition, and to instill a high degree of pride in their own personal appearance, habits and conduct; and to give them real understanding of the meaning and necessity of discipline within a military unit.

[AG 353 (19 Jun 57)]

By Order of *Wilber M. Brucker*, Secretary of the Army:

                        MAXWELL D. TAYLOR,
                     General, United States Army,
Official: *Chief of Staff.*
    HERBERT M. JONES,
Major General, United States Army,
    The Adjutant General.

Distribution:
    Active Army:

CNGB	Sig Bn
Technical Stf, DA	Armor Bn
Admin & Technical Stf Bd	MP Bn
USCONARC	AAA Bn
USARADCOM	Trans Bn
OS Maj Comd	Cml Co
QM Tng Comd	Engr Co
MDW	FA Btry
Armies	Inf Co
Corps	Ord Co
Div	QM Co
Brig	Sig Co
Engr Gp	MP Co
Inf Regt	Armor Co
Armor Gp	AAA Btry
Cml Bn	USMA
Engr Bn	Svc Colleges
FA Bn	Br Svc Sch
Inf Bn	PMST Sr Div Units
Med Bn	PMST Jr Div Units
Ord Bn	PMST Mil Sch Div Units
QM Bn	Mil Mis
	ARMA

NG: State AG; units—same as Active Army.
USAR: Same as Active Army.
For explanation of abbreviations used, see AR 320-50.